SpringerBriefs in Architectural Design and Technology

Series editor

Understanding the complex relationship between design and technology is increasingly critical to the field of Architecture. The SpringerBriefs in Architectural Design and Technology series aims to provide accessible and comprehensive guides for all aspects of current architectural design relating to advances in technology including material science, material technology, structure and form, environmental strategies, building performance and energy, computer simulation and modeling, digital fabrication, and advanced building processes. The series will feature leading international experts from academia and practice who will provide in-depth knowledge on all aspects of integrating architectural design with technical and environmental building solutions towards the challenges of a better world. Provocative and inspirational, each volume in the Series aims to stimulate theoretical and creative advances and question the outcome of technical innovations as well as the far-reaching social, cultural, and environmental challenges that present themselves to architectural design today. Each brief asks why things are as they are, traces the latest trends and provides penetrating, insightful and in-depth views of current topics of architectural design. Springer Briefs in Architectural Design and Technology provides must-have, cutting-edge content that becomes an essential reference for academics, practitioners, and students of Architecture worldwide.

More information about this series at http://www.springer.com/series/13482

Chao Yuan

Urban Wind Environment

Integrated Climate-Sensitive Planning and Design

Chao Yuan
National University of Singapore
Singapore
Singapore

ISSN 2199-580X ISSN 2199-5818 (electronic)
SpringerBriefs in Architectural Design and Technology
ISBN 978-981-10-5450-1 ISBN 978-981-10-5451-8 (eBook)
https://doi.org/10.1007/978-981-10-5451-8

Library of Congress Control Number: 2017964239

Printed on acid-free paper

This Springer imprint is published by the registered company Springer Nature Singapore Pte Ltd.
part of Springer Nature
The registered company address is: 152 Beach Road, #21-01/04 Gateway East, Singapore 189721, Singapore

Foreword by Prof. Edward Ng

In 2003, Hong Kong suffered from the acute respiratory syndrome (SAR). Nobody knew what it was and many lives were lost. In one housing estate, the disease spreads very quickly. It was later diagnosed that it was due to the close proximity of windows of apartments facing each other in a confined air well. The need for a better understanding of the urban wind environment in high-density cities with a congested living environment was deemed necessary.

The Planning Department of the Hong Kong then commissioned a study titled "feasibility for establishment of air ventilation assessment (AVA) system". In 2006, the consultant completed the study and the Hong Kong SAR Government promulgated the Technical Circular on AVA, and added a new section on AVA in the *Hong Kong Planning Standards and Guidelines*. The two documents have since changed the practice of city planning and building design in Hong Kong. Recently, the idea of a better ventilated city is being promoted in China and a few other Asian cities.

This book is timely. I cannot help wishing we had had it in 2003 or before. All is not lost as cities in Asia are still fast urbanizing. Armed with the knowledge of the book, my hope is that they do not make the same mistake that Hong Kong has been making. The key challenges in front of us have been well explained in the opening chapter of the book. On the one hand, there is this unstoppable human desire for development, but on the other hand, there is also a need to consider the carrying capacity of our cities. For planners and architects, this is indeed a difficult balancing act to achieve.

How one should go about doing what needs to be done is systematically laid out in the following few chapters of the book. The story unfolds from the urban to the neighborhood scales, and then to the building scales. The scientific technicalities and the tools needed are illustrated in detail. How urban wind should interface with a better understanding of air pollution and the contribution of urban trees concludes the book.

The book is an interesting read. However, by the time I come to p. 161 of the book, I suddenly realize that I have to stop and wait. I fully appreciate that the study of the wind environment in the high-density city setting is still an emerging field.

I also recognize that much still needs to be researched. It is just that I can be a bit impatient.

However, knowing Dr. Yuan, who was my Ph.D. student, I know I need not have to wait too long. I am sure the human health impact, the cost–benefit evaluation, the consideration of climate change, the atmospheric boundary layers, and so on of the urban wind environment will be further studied and written in the subsequent volumes by Dr. Yuan. I look forward to reading them. In the meantime, putting on my hat as Dr. Yuan's ex-Ph.D. teacher, I should say in my usual "dry" manner: *it is rather good, carry on*.

Sha Tin, Hong Kong
September 2017

Edward Ng
Yao Ling Sun Professor of Architecture
The Chinese University of Hong Kong

Foreword by Prof. Leslie Norford

Perhaps, the most notable aspect of Yuan Chao's admirable book is that the author, initially trained as an architect, has successfully bridged the intellectual gap that separates the urban climate research community and design professionals. This is the culmination of a journey of several years, spanning Dr. Yuan's graduate education, postdoctoral research, and current professorial appointment. The book represents a real triumph, the outcome of a determined effort to serve the practitioners who are charged with improving the livability of the world's increasingly populated cities.

Architects occasionally talk of "agency," in the sense of Webster's definition as "the capacity, condition or state of acting or of exerting power." The famous dome covering Santa Maria del Fiori was designed and constructed by Brunelleschi, trained as a goldsmith. Today, licensed architects are required by law to work with a myriad of professionals to construct buildings and often cede a large role to some of those professionals in the design phase for want of suitable methods that enable suitably accurate early-design explorations.

The gap between architects and engineers often looms wide. Architects have the capacity in early design to envision strategies for passively conditioning buildings but have lacked a means of quantifying performance. Many engineers have mastered ducts and chillers but not building heat balances and airflows under passive conditions. Transdisciplinary teams can fill the gap, but only for projects that attract and can afford them. Recently, software developed to support designer workflows has enabled early-design explorations of building performance by linking CAD designs to separate performance algorithms or software. This software has been developed by engineers and architects with deep appreciation for each other and by those with formal training in both disciplines.

All of which provides a perspective to understand and appreciate Dr. Yuan's manuscript. Urban climatologists, aware of the environmental challenges in today's cities that include heat waves and air pollution, are developing methods and software that are increasingly capable of producing actionable information but are more complex and computationally intensive than many regulators, designers, and planners can accept in day-to-day practice. Professor Yuan has viewed the needs of practitioners as only a designer or planner could do, and has conducted a series of investigations to provide information that is appropriate in accuracy, depth, and required time of calculation.

As someone who has known and collaborated with Dr. Yuan during his graduate studies, postdoctoral training, and faculty appointment, I am astounded at his vision and delighted to see the individual elements united in this book. Urban airflow is his topic, prompted by his years in Hong Kong and first-hand knowledge of that city's efforts, catalyzed by the work of his Ph.D. advisor, Prof. Edward Ng, to define assess airflows and develop ventilation corridors. Dr. Yuan's method of attack is based on physical scales that separate the major sections of the manuscript: urban, neighborhood, and building. A concluding section focuses on urban air quality and the influence of trees on the wind environment.

At urban scale, Dr. Yuan shows that the wind velocity at a given height in a city, normalized by a reference velocity, correlates well with the frontal area density of the buildings averaged over urban grid cells, particularly at the height of building podiums prevalent in Hong Kong. In turn, frontal area density correlates with the ground coverage ratio familiar to and easily obtained by architects. The strength of the second correlation allows architects and planners to adjust GCR to improve the velocity ratio. Working directly with these correlations in Wuhan, Dr. Yuan shows how potential airflow paths can be routed to avoid areas of high frontal area density that produce low wind velocity ratios.

At neighborhood scale, Dr. Yuan regresses velocity ratio calculated from wind tunnel data with a distance-weighted, point definition of frontal area density that he developed, increasing the resolution of his wind maps to several meters. With this in hand, he plots frontal area density, velocity ratio, and annual average wind speed at high resolution, identifying at low computation cost those neighborhoods with poor ventilation and the impact of planned buildings on their immediate surroundings.

Computational fluid dynamics is necessary at building scale but the complexity, simulation time, and post-processing to derive useful results create significant challenges in practice. Dr. Yuan's carefully designed parametric studies produce a wealth of useful observations about the impact of building setback, separation, and permeability on pedestrian-level airflows and the dispersion of pollutants, for example, from a heavily trafficked urban artery.

Finally, Dr. Yuan extends his morphological methods to include the drag force generated by urban trees, which he parameterizes with leaf area index and the shape of the tree canopy, to show the impact of green coverage ratio on wind velocity ratio in high- and low-density urban areas. Designers can use this information to weigh the advantages of trees (shading and evapotranspiration) against decreased airflows.

By gathering his multiyear efforts and framing them cohesively, Dr. Yuan has empowered designers and planners to estimate urban airflows with sufficient precision to make necessary decisions. Much of his work to date has focused on Hong Kong, leaving open further work, by Dr. Yuan or others, to test the universality of his morphometric methods by evaluating them with wind tunnel or field measurements in other cities. His correlations can be ingested easily in CAD and GIS tools, making it easy to incorporate them into current design and planning workflows and simulation methods. As a result, the world's increasing numerous urban dwellers will have more comfortable and healthier living environments.

Cambridge, Massachusetts, USA
December 2017

Prof. Leslie Norford
Massachusetts Institute of Technology

Foreword by Prof. Thomas Schroepfer

The first decade of this century marked the threshold over which half of the world's population now lives in cities. Furthermore, the urban areas of the world are expected to absorb all of the projected population growth in the next few decades while at the same time drawing in rural populations. To support the increasing population densities in our cities, the development of new integrated planning and design paradigms, research methodologies, and implementation processes has become a pressing issue. In this context, the emergence of new modeling and mapping methods and tools for climate-responsive planning and design have led to the powerful adaptation of data to analyze the behavior and form of complex urban environments.

The methods and tools presented in this book allow for the systematic study of the environmental implications of planning and design decisions, creating a laboratory for testing new ideas and making it possible to derive solutions from analysis-driven generation and evolution. The scientific and evidence-based approach to urban wind environments put forward by the author allows for great specificity of environmental responses, enabling urban planners, urban designers, and architects to make informed decisions about their projects on multiple scales. Providing the various stakeholders with the know-how for more control over their respective planning and design processes, the book is an important contribution to the realization of more intelligent future buildings and cities as well as to the achievement of higher standards of environmental sustainability and enhanced liveability for our urban environments.

Singapore
October 2017

Thomas Schroepfer
Series Editor, SpringerBriefs in Architectural
Design and Technology

Acknowledgements

First and foremost, I would like to thank all my Forwards authors, Prof. Edward Ng, Prof. Leslie Norford, and Prof. Thomas Schroepfer, for describing this little book better than I can do, so that a Preface is no longer needed. I am greatly encouraged by their comments on my book, and as a young researcher on this emerging research field, I must carry on.

No words can express how much I am indebted and appreciate my Ph.D. and postdoctoral supervisors, Prof. Edward Ng and Prof. Leslie Norford. I still remember the conversation we had when I met Prof. Ng for the first time. He laughed at me when I told him that I found his information from the web, and said that it is very dangerous to find a Ph.D. supervisor this way. While I totally agree with him, I feel I was a very lucky guy to be mentored by Prof. Ng all these years. He showed me what we are doing is very important, and what he taught me has been my reference guide whenever I feel difficult and lost in my work and life.

I met Prof. Leslie Norford at International Conference of Urban Climate (ICUC) 8 at Dublin, when I noticed a grandpa-looking scholar who was not very sociable but focused greatly on research presentations. Later, I had the privilege to work with him as exchange student and postdoctoral researcher from 2013 to 2015. Prof. Norford became one of my role models. What I learned from him was not only the knowledge on urban aerodynamic properties, but also how a researcher should focus on what he is doing.

I also want to acknowledge Prof. Thomas Schroepfer for encouraging me to publish my research work. He was my supervisor at Singapore during my second-year postdoctoral work. I would like to thank my colleagues at the Chinese University of Hong Kong, Massachusetts Institute of Technology, Singapore University of Technology and Design, and National University of Singapore, where I conducted all the research in this book.

I would also like to thank Ms. Ayu Sukma Adelia and Mr. Zhang Yangyang, my assistants who helped me to put all the chapters together, and did most of the administrative stuff.

I want to acknowledge my family, especially my wife Vivian. Writing this book while working full-time as a new faculty since 2016 required much more time than I expected. This book would not have been completed without Vivian's support, while I was working overtime during weekends and holidays, Vivian was supporting me and tolerating me unconditionally, raising our children and working full-time. Last but not least, I would like to dedicate this book to my 5-year-old son, Edward, and 1-year-old daughter, Hannah. As Prof. Edward Ng always said, the future belongs to the next generation.

About the Book

In the context of urbanization and compact urban living, conventional experience-based planning and design often cannot adequately address the serious environmental issues, such as thermal comfort and air quality. The ultimate goal of this book is to facilitate a paradigm shift from the conventional experience-based ways to a more scientific, evidence-based process of decision-making in both urban planning and architectural design stages. This book introduces novel yet practical modeling and mapping methods, and provides scientific understandings of the urban typologies and wind environment from the urban to building scale through real examples and case studies. The tools provided in this book aid a systematic implementation of environmental information from urban planning to building design by making wind information more accessible to both urban planners and architects, and significantly increase the impact of urban climate information on the practical urban planning and design. This book is a useful reference book to architectural post-graduates, design practitioners and planners, urban climate researchers, as well as policymakers for developing future livable and sustainable cities.

Contents

About the Author

Chao Yuan is an Assistant Professor in the Department of Architecture of National University of Singapore (NUS). His research interests are the climate-sensitive urban planning and design for livable and sustainable cities, focusing on urban aerodynamic properties that is difficult but critical part of high-density urban climate. He has participated in several key policy-level research projects commissioned by Hong Kong Government, as well as actively involved in a few Chinese (e.g., Wuhan and Macau) projects. He developed the frontal area density understanding, which provides an important knowledge linking the built morphology and the city's aerodynamic potentials. This work is now incorporated into Hong Kong and Wuhan's urban climatic research for city planners' references.

Acronyms and Abbreviations

Symbols

A_{b}	Built area in the total lot area (m^2)
A_{c}	Plant canopy area (m^2)
A_{F}	Front areas facing the wind direction of θ (m^2)
A_{front}	Frontal area (m^2)
A_{site}	Site area (m^2)
A_{T}	Total lot area (m^2)
A_{t}	Scanned area (m^2)
$A_{\Delta z,x,y}$	Wind frequency-weighted frontal area at the pixel (x, y), in which x and y are the coordinates (m^2)
$A(\theta)_{\mathrm{proj}(\Delta z)}$	Front areas facing the wind direction of θ for a height increment of ΔZ (m^2)
$\bar{c}$	Normalized concentration
$\langle C_0 \rangle$	Reference emission concentration (PPM)
$\langle C \rangle$	Time-averaged concentration of NO_2 (PPM)
C_{D}	Drag coefficient
$C_{\mathrm{D(building)}}$	Drag coefficient of buildings
$C_{\mathrm{D(tree)}}$	Drag coefficient of trees
$C_{\mathrm{D}(z)}$	Sectional drag coefficient
D	Depth of the domain in CFD simulation (m)
$D_{i,m}$ & $D_{T,i}$	Mass diffusion coefficient
d_{met}	Domain thickness (m)
f	Instantaneous value of variables, such as velocity
$\langle f \rangle$	Mean value
f'	Fluctuating value
h	Canopy height (m)
I	Turbulent intensity
i	Number of species
$\vec{J_i}$	Mass diffusion in turbulent flows

L	Distance (m)
L_y	Mean breadth of the roughness elements facing the wind direction of θ (m)
$l_{x,y}$	Distance coefficient
Max. Lp	Maximum continuous projected façade length (m)
H	Height of the tallest building on site (m)
k	Coefficient related to GCR and λf (0–15 m)
n	Number of the test points
P	Annual probability of winds at a particular direction (%)
P_ratio	Plot ratio
P_i	Annual wind frequency
r	Radius (m)
R_b	Blockage ratio (%)
R_i	Rate of the product from chemical reaction
S_t & S_i	Area of wind tunnel tests and area of the grid in the wind tunnel test area (m^2), respectively
sc_t	Turbulent Schmidt number (0.7)
ΔT_{u_r}	Urban heat island intensity (°C)
t	Time (s)
U	Mean width of street canyon (m)
U_{500}	Annually averaged wind speed at the reference height (m/s)
$U_{x,y,z}$	Wind speed in alongwind, crosswind, and vertical directions, respectively (m/s)
u_t	Turbulent viscosity
u_*	Friction velocity (m/s)
U_b	Input wind velocity of tracer gas at the building height (m/s)
U_c	Averaged wind speed (m/s)
U_{met}	Wind speed at the top of the domain (m/s)
U_p	Pedestrian-level wind speed (m/s)
$U_{reference}$	Wind speed at a reference height (m/s)
U_{ref}	Mean velocity at the reference height (z_{ref})
$\vec{v}$	Overall velocity vector
$V_{\_p}$	Wind speed at the pedestrian level (m s^{-1})
$V_{\_c}$	Wind speed at the top of urban canopy layer (m s^{-1})
$V_{p,i,j}$	Wind velocity at the pedestrian level in a particular wind direction (i) at j-th test point (m s^{-1})
V_h	Wind velocity in the input vertical wind profile (m s^{-1})
$V_{met,i}$	Meteorology data of wind velocity at the reference height d_{met} in a particular wind direction (m/s)
V_{domain}	Volume of the computation domain (m^3)
V_{model}	Volume of the model (m^3)
V_{met}	Wind speed at the top of the domain
$V_{\_s}$	Wind speed at the top of roughness sublayer (m s^{-1})
$V_{\_z,i}$	Mean wind speed at the height of "z" at the wind direction 'I' (m s^{-1})

$\mathrm{VR}_{\_500}$	Average of the directional wind velocity ratios
$V_{\_500,i}$	Mean wind speed at 500 m at the wind direction of i (m s^{-1})
$\mathrm{VR}_{\_500,i,j}$	Directional wind velocity ratio at the wind direction of i
$\mathrm{VR}_{\_w,j}$	Overall wind velocity ratio at j-th test point
$\mathrm{VR}_{i,j}$	Wind velocity ratio in the particular wind direction (i) at j-th test point
W	Width of the domain in CFD simulation (m)
w	Average building width (m)
W_s	Wind velocity of Ethylene (C_4H_4) when it was released from the point source as tracer gas (m/s)
X_1	Height of the test line (Top)
X_2	Height of the test line (Medium)
X_3	Height of the test line (Bottom)
z_d	Zero-plane displacement height (m)
z_{min} & z_{max}	Layer boundary height for logarithmic wind profile (m)
z_0	Surface roughness length for momentum (m)
z_r	Blending height (m)
z_H	Averaged building height (m)
Δz	Height increments in the calculation of $\lambda_{f\ (z)}$ (m)
α	Surface roughness factor
δ	Boundary layer height (m)
κ_0	von Kármán constant
κ	Turbulent kinetic energy (m^2/s^2)
λ_f	Frontal area index
$\lambda_{f(\theta)}$	Frontal area index at the wind direction of θ
$\lambda_{f(z,\theta)}$	Frontal area density at the wind direction of θ
$\lambda_{f(z)}$	Frontal area density accounting for the annual wind probability coming from 16 main directions
$\lambda_{f(0-15\ m)}$	Frontal area density of the podium layer (Δz: 0–15 m)
$\lambda_{f(15-60\ m)}$	Frontal area density of the building layer (Δz: 15–60 m)
$\lambda_{f(0-60\ m)}$	Frontal area density of the urban canopy layer (Δz: 0–60 m)
λ_{f_point}	New point-specific frontal area density
$\lambda_{f\_building}$	Frontal area density of buildings
λ_{f_tree}	Frontal area density of trees
λ_f^*	Frontal area density above the displacement height z_d
λ_f'	Frontal area density below the displacement height z_d
λ_i	Integrated permeability
λ_p	Site/Ground coverage ratio
λ_p'	Permeability of buildings
ω	Specific turbulence dissipation rate (s^{-1})
τ_{tp}	Pressure drag
τ_{ts}	Skin drag
τ_w	Shear stress
ρ	Fluid density (g m^{-3})

ρD	Canopy drag force, or specifically the body force per unit volume on the spatially averaged flow
$\rho\langle\overline{u'w'}\rangle$	Momentum flux caused by turbulent mixing
η	Kolmogorov scales

Subscripts

θ, i	Wind direction
j	Test point
x, y, z	Alongwind, crosswind, and vertical directions, respectively
met	Meteorology data

Abbreviations

AIJ	Architectural Institute of Japan
ASD	Architecture Service Department
AVA	Air Ventilation Assessment
B(P)R	Building (Planning) Regulation
BL	Boundary layer
CALMET	The California Meteorological Model
CFD	Computational Fluid Dynamics
CI	Confidence Intervals
DEM	Digital Elevation Model
DNS	Direct Numerical Simulation
DPA	Development Permission Area
DSP	Development Scheme Plan
EDAs	Existing Development Areas
EEA	European Environment Agency
EPA-ORD	The US EPA's Office of Research and Development
FAD	Front Area Density
GCR	Ground Coverage Ratio
GHS	Greenhouse Gas
GIS	Geographic Information System
GMP	Green Master Planning
GSA	Gross Site Area
HK EPD	Hong Kong Environmental Protection Department
HKBD	Hong Kong Building Department
HKDB	Hong Kong Development Board
HKPD	Hong Kong Planning Department
HKPSG	Hong Kong Planning Standards and Guideline
HKSAR	Hong Kong Special Administrative Region
IENV	Institute for the Environment
IPCC	Intergovernmental Panel on Climate Change
ku_d & ku_0	Kutzbach, J. model

LAI	Leaf Area Index
Le_0	Lettau, H. model
LES	Large Eddy Simulation
LIDAR	Light Detection and Ranging
MDPR	Maximum Domestic Plot Ratio
MM5	Fifth-Generation Penn State/NCAR Mesoscale Model
NCAR	National Center for Atmospheric Research
NDAs	New Development Areas
NDVI	Normalized Difference Vegetation Index
NSA	Net Site Area
OZP	Outline Zoning Plan
PALM	Parallelized Large eddy Simulation (LES) model
PBL	Planetary Boundary Layer
PET	Physiological Equivalent Temperature
PM	Particulate Matter
R	Plant typology ratio between vertical and horizontal crown projected areas
Ra_0	Raupach, M. R., model
RANS	Reynolds-averaged models
RSL	Roughness Sublayer
RSM	Reynolds Stresses Model
RSP	Respirable Particles
SARS	Severe Acute Respiratory Syndrome
SBD	Sustainable Building Design Guideline
SC	Site Coverage Ratio
SE	Standard Error
SST	Shear Stress Transport
SVF	Sky View Factor
TKE	Turbulence Kinetic Energy
UBL	Urban Boundary Layer
UCL	Urban Canopy Layer
UHI	Urban Heat Island
UHII	Urban Heat Island Intensity
US EPA	The United States Environmental Protection Agency
VBA	Visual Basic for Applications
VR	Velocity Ratio
WHO	World Health Organization
WRF	Weather Research and Forecasting model

List of Figures

List of Tables

Chapter 1
High-Density Planning and Challenges

1.1 Current Urban Development

1.1.1 Urbanization

Defined by Davis (1965), urbanization refers to "*the increase over time of the proportion of the total human population that is urban as opposed to rural*". The global urban population has dramatically grown at a faster rate than that of the entire global population (European Environment Agency (EEA) 2010a). Urbanization started in Great Britain in the 1900s as a result of the industrial revolution (Davis 1965). In the 1950s, the rate of urbanization doubled that of the preceding 50 years (Davis 1965). Figure 1.1 shows the urbanization in Hong Kong that was first developed in the 1950s, and current urban populations have reached 100%. Similarly, major developing third world countries, such as those in Southeast Asia and South America, experience a high rate of urbanization in recent decades (LeGates and Stout 2003). While there were only six tropical cities with populations of more than one million in 1940, this number had increased to 52 in 1970 (Glenn and Simon 1998). It is estimated that 60% of the population in Asia, Africa, Latin America, and Oceania will be urbanized by 2025 (Peterson 1984).

1.1.2 Suburbanization (Urban Sprawl)

Urban sprawl, also known as suburbanization, occurs when urban residents move away from the urban areas to the suburbs for better environmental living quality, resulting in the spread of the cities. Urban sprawl is characterized by low-density spatial expansion in the urban fringe (EEA 2006) and specifically occurs in developed regions, such as North America and Europe (Garreau 1992). Due to the healthier lifestyle and the advanced urban services, high-income residents living in

C. Yuan, *Urban Wind Environment*, SpringerBriefs in Architectural Design and Technology, https://doi.org/10.1007/978-981-10-5451-8_1

Fig. 1.1 Urbanization of Hong Kong from 1960s. *Source* Google photos, edited by author

Fig. 1.2 Levittown sprawl. *Source* https://misfitsarchitecture.com

metropolitan districts prefer to live in suburban areas rather than in inner cities or rural areas (EEA 2006). Figure 1.2 shows the sprawl of Levittown, a project nearby New York, started in 1947 as American's prototypical postwar planned community with more than 17,000 detached houses. Figure 1.3 shows the sprawling freeway (18-lane wide) near Toronto, Canada, which is the symptom of the car-dependence community. Urban sprawl is also considered as one of the main reasons for urban environmental and social decay, insufficient resources utilization, and low-income residential areas in inner cities (Bhatta et al. 2010; Garreau 1992).

While the pace of urbanization and city development across developed countries was similar, those at the developing countries differ greatly. As shown in Fig. 1.4, for example, Jakarta was at the stage of suburbanization during the 1990s, with its population density decreasing significantly at the city center, but spreading to the

Fig. 1.3 Sprawling freeway (18-lane wide) near Toronto, Canada, with a suburbanized industry area in the background. *Source* https://en.wikipedia.org/wiki/Suburbanization

suburban areas that were 7–12 km away from the city center (Murakami et al. 2005). In contrast, Metro Manila was at the urbanization stage around the same time, with growing population density at both urban and suburban areas.

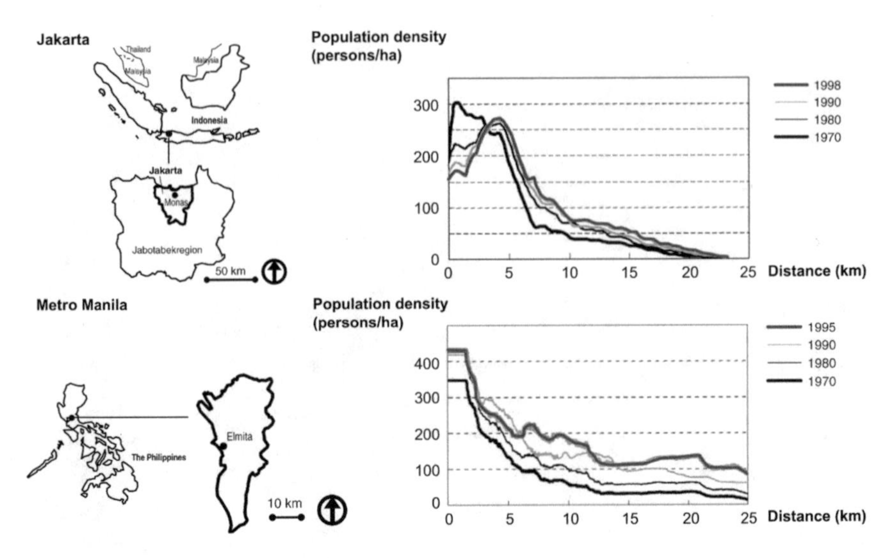

Fig. 1.4 Different urban development stages in Jakarta and Metro Manila. *Source* Murakami et al. (2005), edited by author

Urban sprawl may be associated with increasing urban populations, i.e., urbanization (Murakami et al. 2005). For instance, the Sixth Population Census in Shanghai, China (Statistic of Shanghai 2010) has documented a rapid urbanization in the first decade of the twenty-first century, in which nonlocal population who migrated to Shanghai had grown to 5.5 million in 2010, increased by 159% since 2000, with 6.28 million total urban population increase. However, while the total urban population increased, the population density at the city centers did not necessarily shoot up (Befort et al. 1988). In fact, the population densities at certain old districts even decreased, as shown in Fig. 1.5. Meanwhile, the morphological area in Shanghai has been significantly spreading since 1975. The current urban area is several times larger than that in 1975 when the reform and opening-up in China first began (Zhao et al. 2006).

1.1.3 Negative Effects of Current Urban Development on Urban Climate

A significant and lasting change in the statistical distribution of weather patterns over periods ranging from decades to millions of years has demonstrated the existence of climate change (Solomon et al. 2007). Because of climate change, heat waves are likely to occur more often and last longer, thereby imposing tremendous economic and health burden to the societies (Intergovernmental Panel on Climate Change 2014). The current urban development can exacerbate the adverse impact of

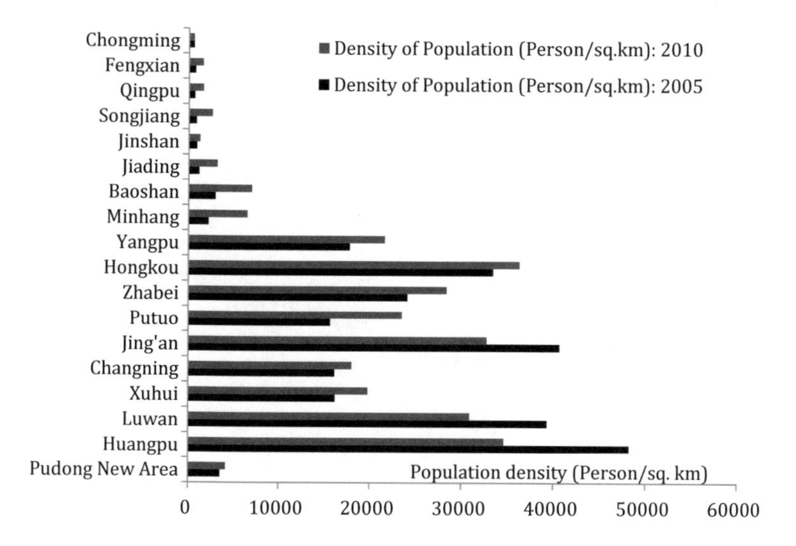

Fig. 1.5 Population densities at different districts of Shanghai in 2005 and 2010. Compared with Baoshan, Minhang, and Putuo that are new districts, the population densities at the old districts, such as Huangpu, Luwan, and Jing'an, significantly decreased due to the urban sprawl. *Source* Statistic of Shanghai (2010)

climate change on people's daily lives (EEA 2010a). Both urbanization and suburbanization increase the Urban Heat Island (UHI) intensity and further intensify the impacts of heat waves in cities. Several studies had reported that a high-temperature urban environment not only increases thermal discomfort but also threatens public health, resulting in high morbidity and mortality (Chan et al. 2010; Goggins et al. 2012; EEA 2010a). The research conducted by Chan et al. (2010) indicated that an average 1 °C elevation above the daily mean air temperature of 28.2 °C (the cutoff temperature point for the Hong Kong population) was associated with 1.8% increase in total mortality. In Europe, similar research has also shown 4% increase in mortality associated with a 1 °C increase in air temperature above the threshold value (EEA 2010a). It should be noted that the association between mortality and heat stress varies by population vulnerability, in which population that is economically deprived tends to be more sensitive to the adverse impact of climate change caused by urbanization and suburbanization (EEA 2010b).

Urban development and urban environment can be linked in a vicious cycle. First, the rapid urbanization increases the consumption of natural resources in order to attain better urban services, which in turn aggregate the negative impacts of the climate change. The ecological footprint of London, for example, is approximately 300 times its geographical area (EEA 2010a). Higher energy consumption (e.g., energy used for cooling, heating, and transportation) in the cities results in greater emission of anthropogenic heat, air pollution, and Greenhouse Gas (GHG). An unfavorable urban

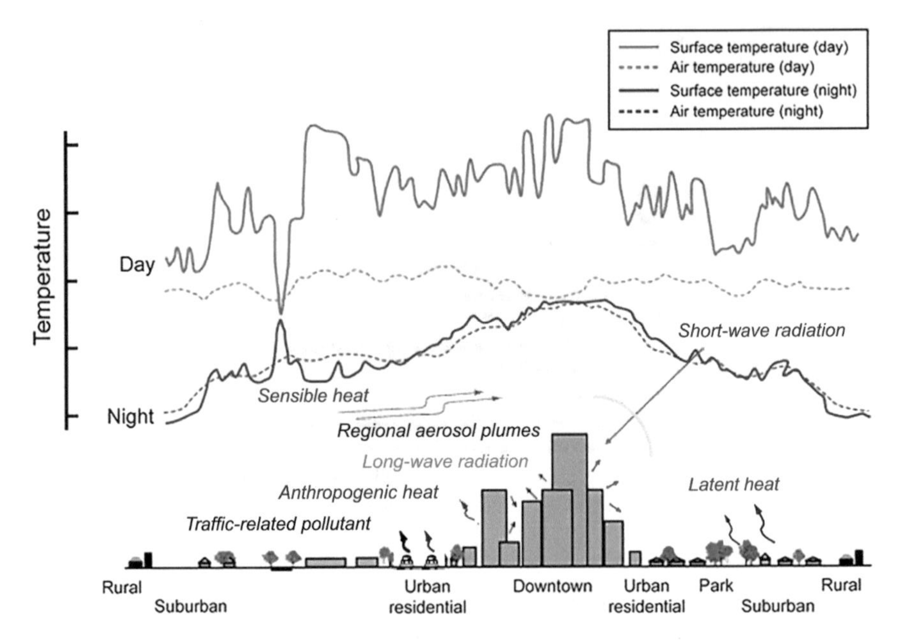

Fig. 1.6 Urban Heat Island and microclimate in built environment. *Source* EEA (2010a), edited by author

climate caused by these anthropogenic emissions further demands more energy and resources to compensate for the negative impacts, and to improve the quality of urban environment. Second, increased occurrence of thermal discomfort and air pollution associated with urbanization promotes suburbanization, i.e., urban sprawl. While urban dwellers mitigate from downtown areas to the urban fringe for a healthier living environment, they often travel back to the city center to work and enjoy better metropolitan services. As a result, such suburbanization movement increases vehicle mileage, which translates into greater traffic energy consumption and emissions. In addition, more land consumption is needed as cities expand to accommodate urban sprawl. As shown in Fig. 1.6, the increase in soil sealing area, poor nocturnal cooling effect, high heat absorbing volume, and less greening at the urban fringe increase the intensity of UHI, and thus worsen the living environment, resulting in continuous movement of urban dwellers to the new urban outskirt. This vicious cycle continuously increases the consumption of energy and other natural resources, intensifies UHI, and exacerbates other existing urban environmental issues (e.g., air pollution), thereby increasing the vulnerability of cities to climate change.

1.2 High-Density Urban Planning

1.2.1 Strengths of High-Density Urban Planning

Since the current urban development undermines the efforts to mitigate the adverse impact of the global climate change, a proactive and proportionate modification to the current urban development is needed. In 2006, the European Environment Agency (EEA) (2006) evaluated two alternative paths of urban development: compact (i.e., high-density) urban development and scattered urban development. Using Madrid as a case study, the land use patterns from 2000 to 2020 in three scenarios with different urban development paths (A: current, B: compact, and C: scattered development) were cross-compared. The findings were shown in Fig. 1.7, which indicates that the compact development significantly decreases land consumption, in which the land consumption for the built-up areas (230.52 km^2) and the loss of agricultural land (84.71 km^2) are only about half of those in the other two scenarios, whereas the scattered development requires more land consumption compared to the current development (EEA 2006). Figure 1.8 shows another study of 32 cities by Newman and Kenworthy (1999), in which the authors found a strong link between urban development densities and petroleum consumption, in that compact urban planning is associated with less energy consumption and less private mobility. This is supported by the fact that compact or high-density living in the cities is generally characterized by shorter journeys to work and services, and greater utilization of walking, cycling or public transport, as well as multifamily houses or blocks that require less heating and less ground space per person. Thus, urban dwellers generally consume less energy and land per capita than rural residents (EEA 2010a).

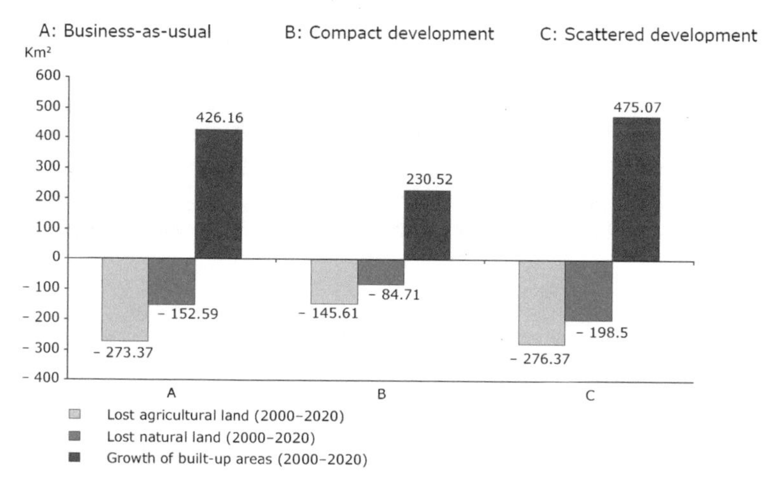

Fig. 1.7 Land use pattern modeling for Madrid by 2020. A: Current development trend, B: compact development, and C: scattered development. *Source* EEA (2006)

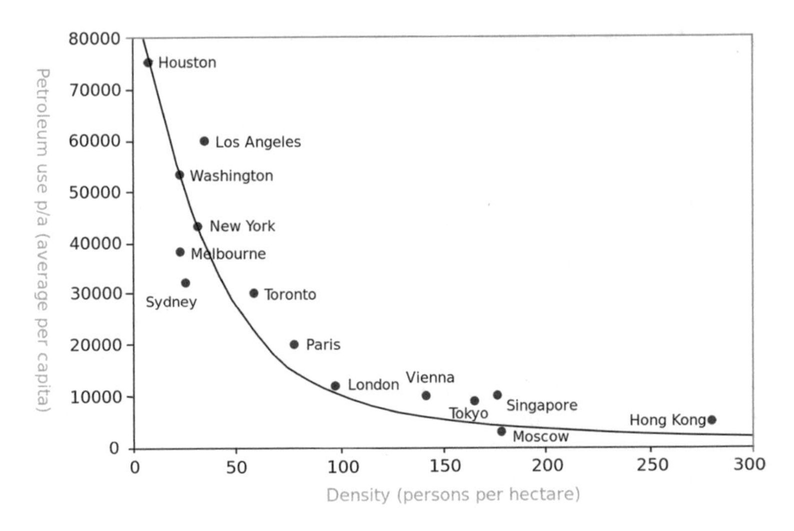

Fig. 1.8 Relationship between transport and urban density. *Source* Newman and Kenworthy (1999)

In terms of energy and land consumption, compact urban planning is considered a good alternative to reverse the vicious cycle between urban development and urban environment, and to prepare the cities to better adapt to climate change. Hong Kong is an excellent example of a high-density city. Given the political restriction of Hong Kong as Special Administrative Region of China, urban sprawl within Hong Kong territory is restricted, except by the limited sea fill reclamation. More importantly, the built-up areas (272 km^2), including new towns and metropolitan areas, are strictly controlled by local urban policies for land use planning (Ng 2009). Such areas only account for about 25% the total territory (1104 km^2) of Hong Kong. Hong Kong has a highly efficient highway system (i.e., no traffic lights situated along highways and well-organized highway exits and entrances) and urban rail transit system (MTR), connecting new towns and metropolitan areas. Therefore, as one of the high-density cities in the world, Hong Kong is a successful exemplary in land utilization and low energy consumption as shown in Figs. 1.8 and 1.9. Figure 1.9 indicates that the energy consumption and carbon dioxide emission in Hong Kong are significantly lower than cities at the United States with similar Human Development Index.

1.2.2 Limitations of High-Density Urban Planning

Despite the apparent strengths of compact urban development, studies have also identified serious environmental issues (e.g., poor ventilation performance and air pollution) associated with compact urban development (Arnfield 2003; Ng 2012; Ng et al. 2011). For example, while the compact and high building blocks at Hong Kong may promote high performance in land and energy usage, such development

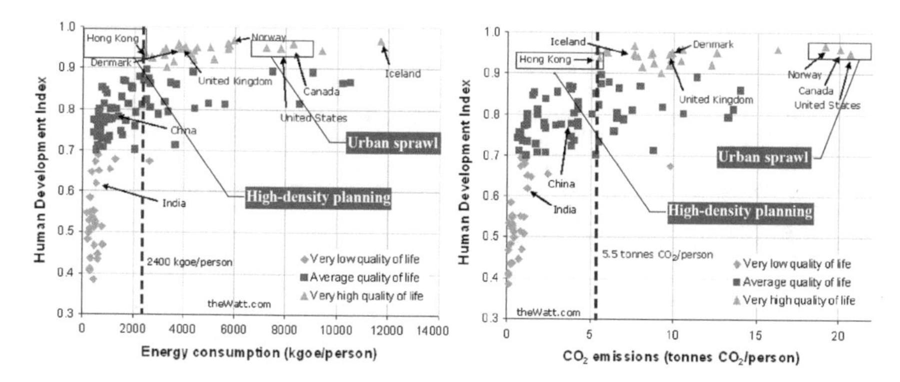

Fig. 1.9 Relationship among Human Development Index, energy consumption per person, and CO_2 emission per person. *Source* United Nations Development Programme. http://hdr.undp.org/en/data

Fig. 1.10 Vehicle fleets in the deep street canyon of Mong Kok and Wan Chai in Hong Kong. *Source* Yuan and Ng (2012)

also leads to stagnant air in the high-density urban areas, as shown in Fig. 1.10. The impact of compact urban development, not climate change, on airflow over the past decades is evident in King's Park monitoring station, monitored by the Hong Kong Observatory (HKO), where the annual mean wind speed measured (i.e., red lines and dots in Fig. 1.11) decreases by 40% from 3.5 to 2.0 m/s at 20 m above the ground (building rooftop) from 1960 to 2015. This is in sharp contrast with stable annual mean wind speed in the Waglan Island station, a reference station that is not surrounded by the built-up areas, over the same period (i.e., blue lines and dots in Fig. 1.11) (Hong Kong Planning Department 2005).

In terms of outdoor thermal comfort, a decrease of wind speed by 0.4–0.5 m/s during summer at sub/tropical cities is equivalent to a 1.0 °C increase in air temperature, while the outdoor thermal comfort under typical summer conditions generally requires a wind speed of 1.5–1.6 m/s (Cheng et al. 2011). These suggest that the

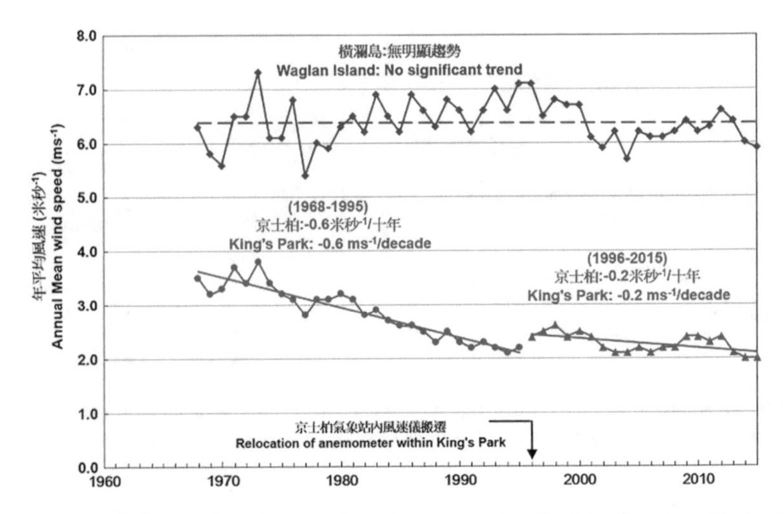

Fig. 1.11 Annual mean wind speed from King's Park and Waglan Island stations. *Source* Hong Kong Planning Department (HKPD) (2005)

decline in average wind speed at the urban areas over the past decades significantly impede the outdoor thermal comfort level. In addition, the reduced wind speed may lead to the accumulation of traffic-related air pollutant at the Urban Canopy Layer (Yuan et al. 2014). As shown in Fig. 1.12, the nitrogen dioxide (NO_2) concentrations measured at the roadside monitoring stations (i.e., the Central, Causeway Bay, and Mong Kok) are much higher than those at the rooftop general monitoring stations, and are also far exceed the threshold values recommended by the World Health Organization (Hong Kong Environmental Protection Department (HK EPD) 2011). While high vehicle density on the streets of high-density urban areas in Hong Kong, as shown in Fig. 1.10, and the new vehicle engine technology may contribute to the higher NO_2 percentage in the total traffic emission documented (EEA 2012; Grice et al. 2009), bulky building blocks and considerably limited open spaces may also prevent traffic air pollutants from dispersing from the deep street canyons (Tominaga and Stathopoulos 2011; Hang et al. 2012). Similar findings were also reported in Europe by the European Environment Agency (EEA) (EEA 2012).

1.3 Objectives and Organization of the Book

A broad literature review has showed that compact or high-density urban development requires less energy and land use than other development options to achieve the same human development under current urban development and climate change atmosphere. As a result, compact urban development is considered as an attractive alternative for the future urban development. The review also indicates that the resultant urban wind environment under the current high-density planning contributes to

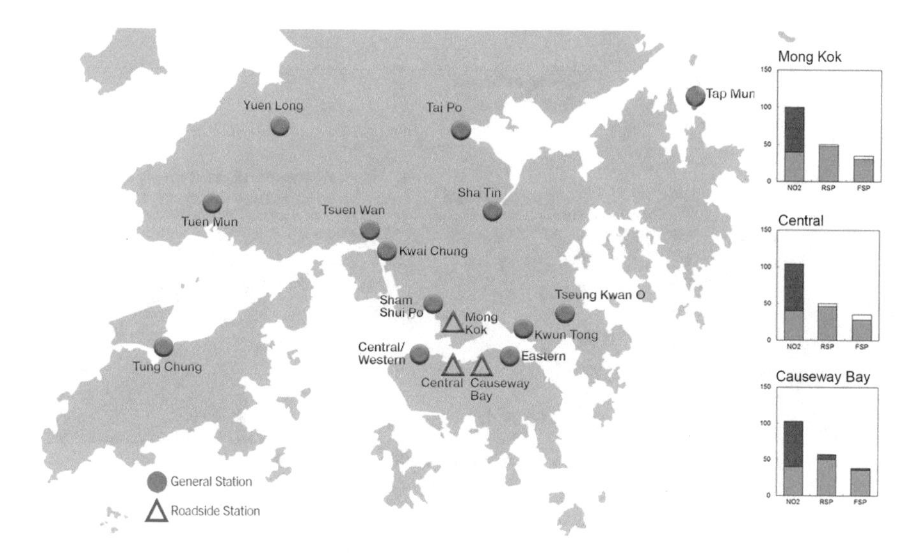

Fig. 1.12 Measured maximum hourly concentrations at Hong Kong, shown in green. The roadside stations have been highlighted as red triangle station, and noncompliance of hourly Air Quality Objective limit values is shown in red. *Source* HK EPD (2011)

negative environmental issues such as poor natural ventilation and air quality, which significantly affect the well-being of urban dwellers. Therefore, climatic knowledge and scientific urban planning and design strategies are essential to make high-density urban areas healthier and more livable.

The study of urban aerodynamic properties normally requires dynamics modeling from regional scale to building scale, where the subjects of urban flow and dispersion with different physical processes vary, as shown in Figs. 1.13 and 1.14. At regional scale, the National Center for Atmospheric Research (NCAR) successively provided two mesoscale prognostic weather models: (1) the Fifth-Generation NCAR/Penn State Mesoscale Model (MM5) (Grell et al. 1994) and (2) the Weather Research and Forecasting Model (WRF) (Skamarock et al. 2005). MM5 has been used in large-scale wind environment studies (Dudhia 1993; Berg and Zhong 2005). Yim et al. (2007) used the MM5 model with the California Meteorological Model (CALMET) model (Scire et al. 2000) to calculate the high-resolution local wind availability in Hong Kong, which is currently an important reference for small-scale studies and urban planning practices. WRF, which is the next-generation prognostic weather model, is currently very popular and is increasingly used in urban environment studies coupled with urban morphology data (Chen et al. 2011; Hong Kong Building Department (HKBD) 2006; Li et al. 2013).

Once the regional modeling is completed, an urban-scale modeling can be conducted to downscale the climatic knowledge, using Computational Fluid Dynamics (CFD) simulation and regional modeling results as input boundary condition. Mellor–Yamada hierarchy model, a higher order turbulence model for atmospheric

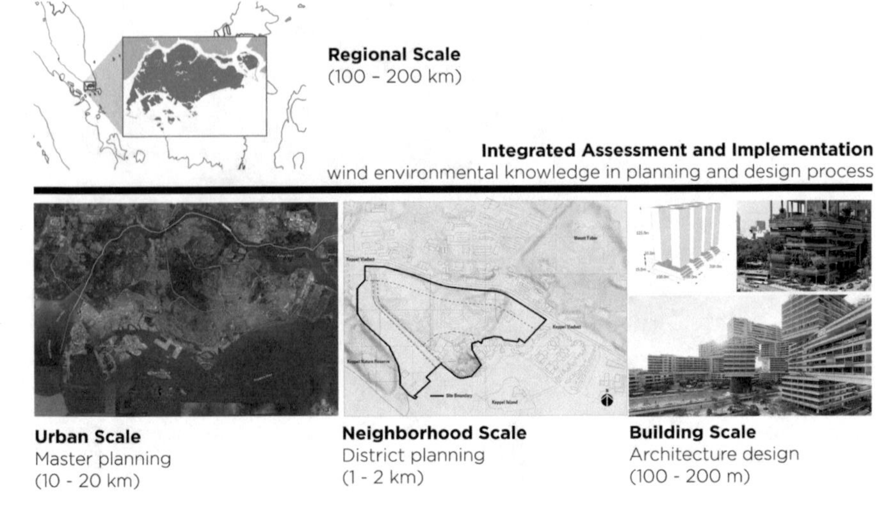

Fig. 1.13 Subjects of urban flow and dispersion at four scales: regional scale for integrated assessment and implementation (Britter and Hanna 2003). *Source* Author

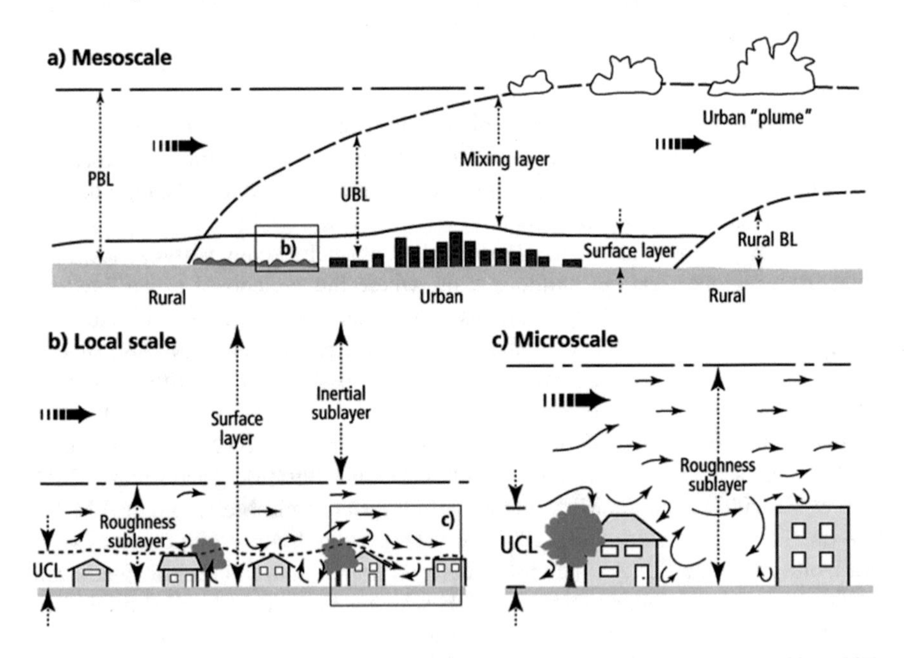

Fig. 1.14 Various physical processes at mesoscale, local scale, and microscale. *Source* Oke (1987)

circulation, was developed to examine the effects of land use on urban environments (Mochida et al. 1997; Murakami 2004). Mochida et al. (1997) and Ashie et al. (2009) conducted urban CFD simulation of Tokyo to understand effects of building blocks on the thermal environment of Tokyo. Modeling results can be further downscaled to neighborhood and even building scale, using CFD simulation with higher resolution. Kondo et al. (2006) simulated NO*x* diffusion at the most polluted roadside areas in Japan. Letzel et al. (2008) studied the characteristics of urban turbulence using the urban version of Parallelized Large Eddy Simulation (LES) model (PALM), which is superior to the conventional Reynolds-averaged models (RANS). Blocken et al. (2007) evaluated the air flow in the gap between two parallel buildings arranged across the wind direction using the commercial CFD code, and the study indicates that the pedestrian-level wind speed is accelerated in the passages between two parallel buildings, known as the channel effect. Hang et al. (2012) conducted a study on the effects of varying building heights on street-level air pollutant dispersion.

The above modeling methods provide important knowledge on wind environment from regional scale to building scale. However, conducting these modeling tests for a particular urban planning or building design is expensive and time-consuming, especially for the regional and urban scales, and the modeling results cannot keep up with the quick planning processes. Therefore, these models have been used mostly as research tool, rather than planning/design tool. More importantly, the scientific uncertainty in the smaller scale modeling would be increased if there are no larger scale modeling results. The urban wind environment understandings linking urban scale through building scale would also be incomplete, making it difficult to develop practical wind environment-sensitive planning and design strategies for different spatial scales, and interweave them into corresponding planning and design stages.

Given the abovementioned growing concerns about the urban wind environment and current gap between existing modeling methods and planning/design practice, this book is designed to provide systematic and practical modeling solutions from urban scale to building scale to address urban natural ventilation and air quality issues. Specifically, practical morphological modeling methods are developed for the wind assessment, and computational parametric studies are conducted to provide the scientific understandings on airflow and pollutant dispersion. Corresponding planning and design strategies are introduced, respectively. This book is organized as four parts: urban-scale wind environment (Part I); neighborhood-scale wind environment (Part II); building-scale wind environment (Part III); and other wind environmental issues (Part IV). The ultimate goal of this book is to reduce the scientific uncertainties in the climate-sensitive planning and design.

References

Arnfield AJ (2003) Two decades of urban climate research: a review of turbulence, exchanges of energy and water, and the urban heat island. Int J Climatol 23(1):1–26

Ashie Y, Hirano K, Kono T (2009) Effects of sea breeze on thermal environment as a measure against Tokyo's urban heat island. Paper presented at the seventh international conference on urban climate, Yokohama, Japan

Befort WA, Luloff AE, Morrone M (1988) Rural land-use and demographic change in a rapidly urbanizing environment. Landscape Urban Plann 16(4):345–356

Berg LK, Zhong S (2005) Sensitivity of MM5-simulated boundary layer characteristics to turbulence parameterizations. J Appl Meteorol 44:1467
Bhatta B, Saraswati S, Bandyopadhyay D (2010) Urban sprawl measurement from remote sensing data. Appl Geogr 30(4):731–740
Blocken B, Carmeliet J, Stathopoulos T (2007) CFD evaluation of wind speed conditions in passages between parallel buildings—effect of wall-function roughness modifications for the atmospheric boundary layer flow. J Wind Eng Ind Aerodyn 95(2007):941–962
Britter RE, Hanna SR (2003) Flow and dispersion in urban areas. Annu Rev Fluid Mech 35:469–496
Chan EYY, Goggins WB, Kim JJ, Griffiths SM (2010) A study of intracity variation of temperature-related mortality and socioeconomic status among the Chinese population in Hong Kong. J Epidemiol Community Health 66(4):322–327
Chen F, Kusaka H, Bornstein R, Ching J, Grimmond CSB, Grossman-Clarke S, Loridan T, Manning KW, Martilli A, Miao SG, Sailor D, Salamanca FP, Taha H, Tewari M, Wang XM, Wyszogrodzki AA, Zhang CL (2011) The integrated WRF/urban modelling system: development, evaluation, and applications to urban environmental problems. Int J Climatol 31(2):273–288
Cheng V, Ng E, Chan C, Givoni B (2011) Outdoor thermal comfort study in a subtropical climate: a longitudinal study based in Hong Kong. Int J Biometeorol 56(1):43–56
Davis K (1965) The urbanization of the human population. Sci Am 213(3):40–53
Dudhia J (1993) A nonhydrostatic version of the Penn State-NCAR mesoscale model: validation tests and simulation of an Atlantic cyclone and cold front. Mon Weather Rev 121:1493–1513
European Environment Agency (EEA) (2006) Urban sprawl in European: the ignored challenge. European Environment Agency, Copenhagen
European Environment Agency (EEA) (2010a) The European environment—state and outlook 2010: urban environment. European Environment Agency, Copenhagen
European Environment Agency (EEA) (2010b) European Union emission inventory report 1998–2008 under the UNECE convention on Long-range Transboundary Air Pollution (LRTAP). European Environment Agency, Copenhagen
European Environment Agency (EEA) (2012) Air quality in Europe—2012 report. Denmark, Copenhagen
Garreau J (1992) Edge city: life on the new frontier. Anchor Books, New York
Glenn RM, Simon N (1998) Tropical climatology, 2nd edn. Wiley, England
Goggins WB, Chan E, Ng E, Ren C, Chen L (2012) Effect modification of the association between short-term meteorological factors and mortality by urban heat islands in Hong Kong. PLoS ONE 7(6):e38551
Grell GA, Dudhia J, Staufeers DR (1994) A descrption of the fifth generation Penn State/NCAR moesoscale model (MM5). National Center for Atmospheric Research
Grice S, Stedman J, Kent A, Hobson M, Norris J, Abbott J, Cooke S (2009) Recent trends and projections of primary NO_2 emissions in Europe. Atmos Environ 43(13):2154–2167
Hang J, Li YG, Sandberg M, Buccolieri R, Di Sabatino S (2012) The influence of building height variability on pollutant dispersion and pedestrian ventilation in idealized high-rise urban areas. Build Environ 56:346–360
Hong Kong Building Department (HKBD) (2006) Sustainable building design guidelines. Practical note for authorized persons, registered structure engineers and registered geotechnical engineers. APP-152, Hong Kong
Hong Kong Environmental Protection Department (HK EPD) (2011) Air quality in Hong Kong 2011—a report on the results from the air quality monitoring network (AQMN). The Goveronment of the Hong Kong Special Administrative Region, Hong Kong
Hong Kong Planning Department (HKPD) (2005) Feasibility study for establishment of air ventilation assessment system, final report, the government of the Hong Kong special administrative region. Hong Kong planning Department, Hong Kong
Intergovernmental Panel on Climate Change (2014) Climate change 2014: synthesis report. Intergovernmental Panel on Climate Change, Geneva

Kondo H, Asahi K, Tomizuka T, Suzuki M (2006) Numerical analysis of diffusion around a suspended expressway by a multi-scale CFD model. Atmos Environ 40(16):2852–2859
LeGates RT, Stout F (2003) The city reader. Urban reader series, 3 edn. Routledge/Taylor & Francis Group, London/New York
Letzel MO, Krane M, Raasch S (2008) High resolution urban large-eddy simulation studies from street canyon to neighbourhood scale. Atmos Environ 42(38):8770–8784
Li XX, Koh TY, Entekhabi D, Roth M, Panda J, Norford L (2013) A multi-resolution ensemble study of a tropical urban environment and its interactions with the background regional atmosphere. J Geophys Res 118(17):9804–9818
Mochida A, Murakami S, Ojima T, Kim S, Ooka R, Sugiyama H (1997) CFD analysis of mesoscale climate in the Greater Tokyo area. J Wind Eng Ind Aerodyn 67–68:459–477
Murakami S (2004) Indoor/outdoor climate design by CFD based on the software platform. Int J Heat Fluid Flow 25:849–863
Murakami A, Zain AM, Takeuchi K, Tsunekawa A, Yokota S (2005) Trends in urbanization and patterns of land use in the Asian mega cities Jakarta, Bangkok, and Metro Manila. Landscape Urban Plann 70:251–259
Newman P, Kenworthy J (1999) Sustainability and cities, overcoming automobile dependence. Island press, Washington D.C.
Ng E (2009) Policies and technical guidelines for urban planning of high-density cities—air ventilation assessment (AVA) of Hong Kong. Build Environ 44(7):1478–1488
Ng E (2012) Towards a planning and practical understanding for the need of meteorological and climatic information for the design of high density cities—a case based study of Hong Kong. Int J Climatol 32:582–598
Ng E, Yuan C, Chen L, Ren C, Fung JCH (2011) Improving the wind environment in high-density cities by understanding urban morphology and surface roughness: a study in Hong Kong. Landscape Urban Plann 101(1):59–74
Oke TR (1987) Boundary layer climates, 2nd edn. Methuen, Inc., USA
Peterson J (1984) Global population projections through the 21st century: a scenario for this issue. Ambio 13:134–141
Scire JS, Robe FR, Fernau ME, Yamartino RJ (2000) A user's guide for the CALMET meteorological model (Version 5). Earth Tech, Inc., Concord
Skamarock WC, Klemp JB, Dudhia J, Gill DO, Barker DM, Wang W, Powers JG (2005) A description of the advanced research WRF version 2. NCAR Technical note. National Center for Atmospheric Research
Solomon S, Qin D, Manning M, Chen Z, Marquis M, Averyt KB, Tignor M, Miller HL (2007) Contribution of working group I to the fourth assessment report of the intergovernmental panel on climate change. Climate Change 2007: the physical science basis. Cambridge University Press, New York
Statistic of Shanghai (2010) Shanghai statistical yearbook. Shanghai Statistics, Shanghai
Tominaga Y, Stathopoulos T (2011) CFD modeling of pollution dispersion in building array: evaluation of turbulent scalar flux modeling in RANS model using LES results. J Wind Eng Ind Aerodyn 104:484–491
Yim SHL, Fung JCH, Lau AKH, Kot SC (2007) Developing a high-resolution wind map for a complex terrain with a coupled MM5/CALMET system. J Geophys Res 112:D05106
Yuan C, Ng E (2012) Building porosity for better urban ventilation in high-density cities—a computational parametric study. Build Environ 50:176–189
Yuan C, Ng E, Norford LK (2014) Improving air quality in high-density cities by understanding the relationship between air pollutant dispersion and urban morphologies. Build Environ 71:245–258
Zhao SQ, Da LJ, Tang HF, Kun S, Fang JY (2006) Ecological consequences of rapid urban expansion: Shanghai, China. Front Ecol Environ 4(7):341–346

Part I
Urban Scale Wind Environment

Chapter 2
Empirical Morphological Model to Evaluate Urban Wind Permeability in High-Density Cities

2.1 Introduction

2.1.1 Background

Hong Kong has one of the highest densities among megacities in the world. Seven and a half million inhabitants live on a group of islands that total 1000 km^2. Hong Kong has a hilly topography; hence, only 25% of the land is built-up areas (Ng 2009). Land prices in Hong Kong have been increasing over the years. For example, in the Central Business District, rent prices have increased 33% between 2005 and 2007 (Hong Kong Rating and Valuation Department (HKRVD) 2009). Owing to the limited land area and the increasing land prices, property developers are building taller and bulkier buildings with higher building plot ratios that occupy the entire site area in order to economically cope with the high land costs, as shown in Fig. 2.1. In addition, the Government of Hong Kong has planned the need to deal with an increasing population, which is projected to increase to 10 million in the next 30 years. Seeking ways to optimize the urban morphology of the city is a difficult but important task for urban planners.

Tall and bulky high-rise building blocks with very limited open spaces in between, uniform building heights, and large podium structures have collectively led to lower permeability for urban air ventilation at the pedestrian level (Ng 2009). The mean wind speeds recorded by the urban observatory stations in urban areas over the past decade have decreased by over 40% (Hong Kong Planning Department (HKPD)

Originally published in Edward Ng, Chao Yuan, Liang Chen, Chao Ren and Jimmy C. H. Fung 2011. Improving the wind environment in high-density cities by understanding urban morphology and surface roughness: a study in Hong Kong. Landscape and Urban Planning 101 (1), pp. 59–74, © Elsevier, https://doi.org/10.1016/j.landurbplan.2011.01.004.

C. Yuan, *Urban Wind Environment*, SpringerBriefs in Architectural Design and Technology, https://doi.org/10.1007/978-981-10-5451-8_2

Fig. 2.1 An urban skyline of Hong Kong

2005). Stagnant air in urban areas has caused, among other issues, outdoor urban thermal comfort problems during the hot and humid summer months in Hong Kong. Stagnant air has also worsened urban air pollution by restricting dispersion in street canyon with high building-height-to-street-width ratios. The Hong Kong Environmental Protection Department (HK EPD) has reported the frequent occurrence of high concentrations of pollutants, such as NO_2 and respirable particles (RSP) in urban areas such as Mong Kok and Causeway Bay (Yim et al. 2009). These areas also have some of the highest urban population densities in Hong Kong.

The 2003 outbreak of the Severe Acute Respiratory Syndrome (SARS) epidemic in Hong Kong had brought attention to how environmental factors (i.e., air ventilation and dispersion in buildings) played an important role in the transmission of SARS and other viruses. Since the outbreak, the planning community in Hong Kong started to pay more attention to the urban design process in order to optimize the benefits of the local wind environment for urban air ventilation. As a result, the Hong Kong Government had commissioned a number of studies on this regard; the most important project among the government-commissioned studies is entitled "Feasibility Study for Establishment of Air Ventilation Assessment System" (AVA), which began in 2003 (Ng 2009). The primary purpose of this comprehensive chapter is to establish the protocol that assesses the effects of major planning and development projects on urban ventilation in Hong Kong (Ng 2007).

The importance of wind environment on the heat, mass, and momentum exchange between urban canopy layer and boundary layer has been studied by urban climate researchers (Arnfield 2003). Two modeling methods have been frequently applied to study wind environment of the city: wind tunnel tests and computational fluid dynamics (CFD) techniques. The United States Environmental Protection Agency (US EPA) conducted numerous urban-scale wind tunnel tests to understand the dispersion of particulate matters smaller than 10 μm in aerodynamic diameter (PM_{10}) (Ranade et al. 1990). Williams and Wardlaw (1992) conducted a large-scale wind tunnel study to describe the pedestrian-level wind environment in the city of Ottawa, Canada, and identified areas of concern for planners. Plate (1999) developed the boundary-layer wind tunnel studies to analyze urban atmospheric conditions, including wind forces on buildings, pedestrian comfort, and diffusion processes from point-sources of the city. Kastner-Klein et al. (2001) analyzed the interaction between wind turbulence and the effects induced by vehicles moving inside the urban canopy. Wind velocity and turbulence scales throughout the street canyons of the city were analyzed using smoke visualization (Perry et al. 2004). In 2004, the US EPA's Office of Research and Development (EPA-ORD) conducted a city-scale wind tunnel study to analyze the airflow and pollutant dispersion in the Manhattan area (Perry et al. 2004). Kubota et al. (2008) conducted wind tunnel tests and revealed the relationship between plan area fraction (λ_P) and the mean wind velocity ratio at the pedestrian level in residential neighborhoods of major Japan cities. In Hong Kong, the Wind/Wave Tunnel Facility has conducted numerous tests at the city, district, and urban scale to understand the wind availability and flow characteristics of Hong Kong (HKPD 2008).

Apart from wind tunnels, CFD model simulation can be applied at the initial urban planning stage in providing a "qualitative impression" of the wind environment. Mochida et al. (1997) conducted a CFD study to analyze the mesoscale climate in the Greater Tokyo area. Murakami et al. (1999) used CFD simulations to analyze the wind environment at the urban scale. Kondo et al. (2006) used CFD simulations to analyze the diffusion of NO_x at the most polluted roadside areas around the Ikegami-Shinmachi crossroads in Japan. Letzel et al. (2008) conducted studies of urban turbulence characteristics using the urban version of the parallelized large eddy simulation (LES) model (PALM), which is superior to the conventional Reynolds-averaged models (RANS). Using the Earth Simulator, Ashie et al. (2009) conducted the largest urban CFD simulation of Tokyo to understand the effects of building blocks on the thermal environment of Tokyo. Ashie et al. noted that the air temperatures around Ginza and JR Shimbashi are much higher than in the surrounding areas of Hama Park and Sumida River. Ashie et al. argued that the high air temperature can be attributed to the bulky buildings at Ginza and JR Shimbashi that obstruct the incoming sea breezes (Ashie et al. 2009). Yim et al. (2009) used CFD simulation to investigate the air pollution dispersion in a typical Hong Kong urban morphology. In general, using CFD for urban-scale investigation has been gaining momentum in the scientific circle. Two important documents that provide guidelines for CFD usage have been published: Architectural Institute of Japan (AIJ) Guidebook (AIJ 2007; Tominaga et al. 2008) and COST action C14 (Frank 2006).

2.1.2 *Objectives and Needs of This Study*

While the application of wind tunnels and CFD model simulations to analyze the interaction between the urban area and the atmosphere has made important contribution to the understanding of urban air ventilation of the city, such applications are costly, and may not be able to keep up with the fast design process in the initial stages of the design and planning decision-making process. Instead, the outlined and district-based information based on urban morphological data parametrically understood may be more useful for planners.

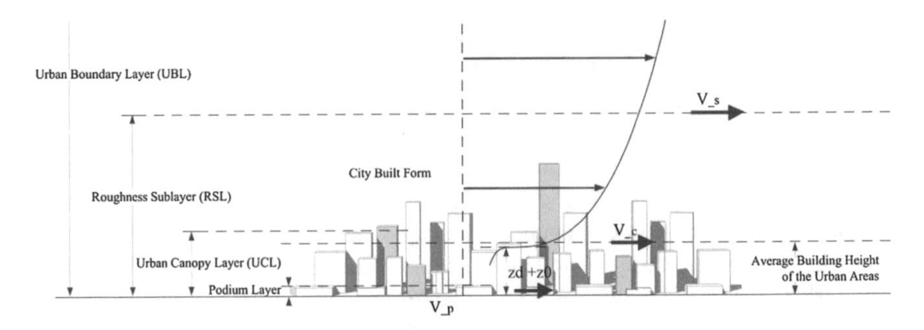

Fig. 2.2 Wind speed profile, podium layer, urban canopy layer, and roughness sublayer. $V_{\_p}$: wind speed at the pedestrian level (2 m above the ground); $V_{\_c}$: wind speed at the top of urban canopy layer; $V_{\_s}$: wind speed at the top of roughness sublayer. The wind speed profile was drawn in accordance with the power law expression: $V_{\_z,i}/V_{\_500,i} = (Z/500)^{\alpha}$, ($\alpha = 0.35$)

This chapter employs the understandings of urban surface roughness to establish the relationship between heterogeneous urban morphologies and urban air ventilation environment. A new method with cross-section areas, which takes into account the site-specific wind information measured at 60 m height using the MM5/CALMET model simulation, was used to calculate the frontal area density (λ_f). Using the site-specific wind information, the new calculation method of λ_f focuses on the effects of the built environment to the wind field, which provides a spatially averaged understanding of wind permeability at the urban scale.

In addition, the new calculation method of λ_f considers the unique urban morphology of the podiums and towers in Hong Kong (i.e., many tall and slender buildings stand on large podiums), which shows that taking the urban morphology of podiums into consideration is important. Therefore, the podium layer is defined within the urban canopy layer as shown in Fig. 2.2. The spatial characteristics of the large podiums provide much larger drag force on airflow nearer to the ground than the upper layers, and thus can greatly affect the wind environment at the pedestrian level.

This study first correlates the pedestrian-level wind environment with λ_f calculated at the podium layer, and then establishes an understanding of surface roughness and urban morphology based on ground coverage ratio (GCR), a term familiar to urban planners, with λ_f to simplify the practical application of the understanding for professional use.

2.2 Literature Review

2.2.1 Roughness Characteristics

The roughness properties of urban areas affect surface drag, scales and intensity of turbulence, wind speed, and the wind profile in urban areas (Landsberg 1981). The total drag on a roughness surface includes both a pressure drag (τ_{tp}) on the roughness elements and a skin drag (τ_{ts}) on the underlying surface (Shao and Yang 2005). In this study, only the pressure drag is considered, since skin drag is relatively small and is not a factor that can be controlled at the urban scale. Oke (1987) provided the logarithmic wind profile in a thermally neutral atmosphere, which is a semiempirical relationship that acts as a function of two aerodynamic characteristics: roughness length (z_0) and the zero-plane displacement height (z_d). The reliable evaluation of such aerodynamic characteristics of urban areas is significant in depicting and predicting urban wind behaviors (Grimmond and Oke 1999).

Currently, three methods can be used to estimate the surface roughness: Davenport roughness classification (Davenport et al. 2000), morphologic, and micrometeorological methods (Grimmond and Oke 1999). The Davenport Classification is a surface-type classification based on the assorted surface roughness values, using high-quality observations (Davenport et al. 2000), which covers a wide range of surface types. This method is not too helpful to describe urban permeability in high-density cities, because most of the urban areas could only be described in Class 8 "Skimming: City centre ($z_0 \geq 2$)". Compared with the micrometeorological method, the morphometric method estimates the aerodynamic characteristics, such as z_0 and z_d, using empirical equations (Lettau 1969; MacDonald et al. 1998; Raupach 1992; Bottema 1996; Kutzbach 1961). Grimmond and Oke (1999) validated the empirical models by Kutzbach, Lettau, Raupach, Bottema, and Macdonald. While reasonable relationships between z_0 and frontal area index ($\lambda_{f(\theta)}$) for low- and medium density forms have been found, there is a tendency of overestimation of z_0 for higher density cases (Bottema 1996).

Grimmond and Oke (1999) calculated $\lambda_{f(\theta)}$ in the context of the urban morphology of North America cities. Ratti et al. (2002) calculated $\lambda_{f(\theta)}$ of 36 wind directions in London, Toulouse, Berlin, and Salt Lake City. By incorporating a spatially continuous database on aerodynamic and morphometric characteristics, such as $\lambda_{f(\theta)}$, z_0, and z_d, morphometric estimation methods can be helpful to urban planners and researchers in depicting the distribution of the roughness of the city. Using Bottema's model equation, Gál and Unger (2009) mapped z_0 and z_d to detect the ventilation paths in Szeged. Wong et al. (2010) mapped $\lambda_{f(\theta)}$ to detect the air paths in the Kowloon Peninsula of Hong Kong.

2.2.2 *Calculation of Frontal Area Index and Frontal Area Density*

The frontal area index $\lambda_{f(\theta)}$ is a function of wind direction of θ, which is an important parameter of the wind environment. The $\lambda_{f(\theta)}$ in a particular wind direction of θ is defined (Raupach 1992) as

$$\lambda_{f(\theta)} = \frac{A_F}{A_T} - L_y \cdot Z_H \cdot \rho_{el}, \tag{2.1}$$

where A_F represents the front areas of buildings that face the wind direction of θ, A_T represents the total lot area, L_y represents the mean breadth of the roughness elements that face the wind direction of θ, Z_H represents the mean building height, and ρ_{el} represents the density (number) of buildings per unit area. $\lambda_{f(\theta)}$ has been used widely by researchers in plant canopy and urban canopy communities to help quantify drag force.

Frontal area density, $\lambda_{f(z,\theta)}$, represents the density of $\lambda_{f(\theta)}$ at a height increment of "z" (Burian et al. 2002):

$$\lambda_{f(z,\theta)} = \frac{A(\theta)_{proj(\Delta z)}}{A_T}, \quad (2.2)$$

where $A(\theta)_{proj(\Delta z)}$ represents the area of building surfaces that approaches a wind direction of θ for a specified height increment "Δz" and A_T represents the total lot area of the study area.

Compared with $\lambda_{f(\theta)}$, which is an average value that describes the urban morphology of the entire urban canopy, $\lambda_{f(z,\theta)}$ represents a density that describes the urban morphology in the interested height band. Burian et al. (2002) conducted frontal area density calculations in a height increment of 1 m in Phoenix City, and found that $\lambda_{f(z,\theta)}$ is a function of land uses because the buildings in different land uses have different building morphologies.

2.3 Development of New Layer-Based $\lambda_{f(z)}$

Due to the morphological difference between the podium layer and building layer in Hong Kong, as shown in Fig. 2.2, the respective $\lambda_{f(z,\theta)}$ of the layer is expected to be better than $\lambda_{f(\theta)}$ in capturing and describing the complicated urban morphology in Hong Kong. Using a high-resolution (1 × 1 m) three-dimensional building database with building height information and digital elevation model (DEM), a self-developed program embedded as a VBA script in the ArcGIS system was applied to calculate the frontal area density ($\lambda_{f(z)}$) at different height bands. $\lambda_{f(z)}$ accounts for the annual wind probability from 16 main directions:

$$\lambda_{f(z)} = \sum_{\theta=1}^{16} \lambda_{f(z,\theta)} \cdot P_\theta, \quad (2.3)$$

where $\lambda_{f(z,\theta)}$ represents the frontal area density at a particular wind direction (θ), and can be calculated with Eq. 2.2. P_θ represents annual wind probability at a particular direction (θ).

2.3.1 *Height of the Podium and Urban Canopy Layer*

To identify the height of the podium and urban canopy layer in high-density urban areas of Hong Kong, a statistical study was conducted based on three-dimensional building database provided by the Hong Kong Government. Twenty-five urban areas have been sampled as shown in Fig. 2.3. Mean and upper quartiles of the building and podium height at metropolitan and new town areas were calculated. According to the height distribution shown in Fig. 2.4, the urban canopy layer and podium layer at the metropolitan areas were identified as 60 and 15 m, respectively. Figure 2.5 shows the calculated $\lambda_{f(0\text{–}15\ m)}$, $\lambda_{f(15\text{–}60\ m)}$, and $\lambda_{f(0\text{–}60\ m)}$, corresponding to the height increments of 0–15 m (podium layer), 15–60 m (building layer), and 0–60 m (urban canopy layer), respectively.

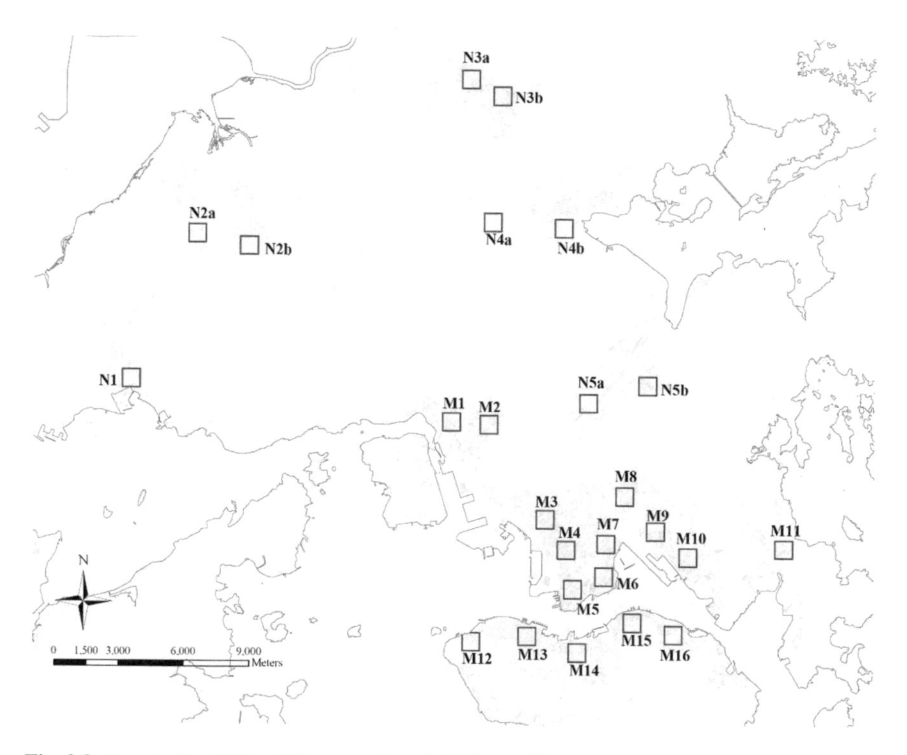

Fig. 2.3 Twenty-five 900 × 900 m test sites (M: Metropolitan areas; N: New town areas)

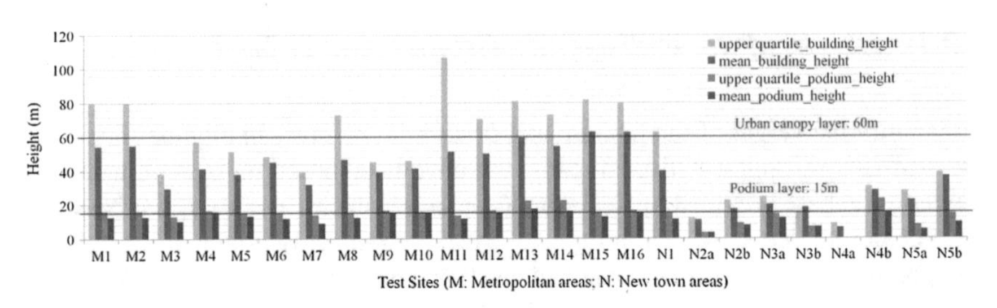

Fig. 2.4 Heights of the urban canopy layer and the podium layer. Based on the understanding, the heights of the urban canopy layer and the podium layer are set in 60 and 15 m for this chapter, respectively

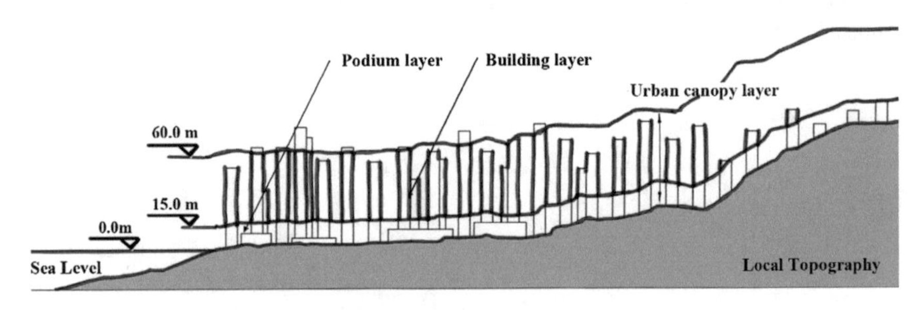

Fig. 2.5 Illustration of the building layer, podium layer and urban canopy layer

2.3.2 Wind Availability in Hong Kong (MM5/CALMET System)

In Hong Kong, the local topography and land–sea contrast impose significant influence on the wind direction in the immediate vicinity of the urban canopy layer, as shown in Fig. 2.5. Therefore, to focus on the impact of building drag force on airflow, site-specific wind roses for annual non-typhoon winds at 60 m height in 16 directions were used to calculate the corresponding local values of $\lambda_{f(z)}$. Due to the complex topography of Hong Kong, the territory is divided into subareas, with various area-specific wind roses (Fig. 2.6). The data on site-specific wind roses were obtained from the fifth-generation NCAR/PSU mesoscale model (MM5) that incorporates the California Meteorological (CALMET) system (Yim et al. 2007). MM5 is a limited-area, non-hydrostatic, and terrain-following mesoscale meteorological model. MM5 is designed to simulate mesoscale and regional-scale atmospheric circulation (Dudhia 1993; Yim et al. 2007). CALMET is a diagnostic three-dimensional meteorological model that can interface with MM5 (Scire et al. 2000).

The terrain in Hong Kong is complex; hence, the resolution used in MM5 simulations (typically down to 1 km) cannot accurately capture the influence of topology characteristics on wind environment. Therefore, CALMET, a prognostic meteorological model capable of higher resolutions (down to 100 m), was used. Combining the data obtained from MM5 and the data obtained from an upper air sounding station,

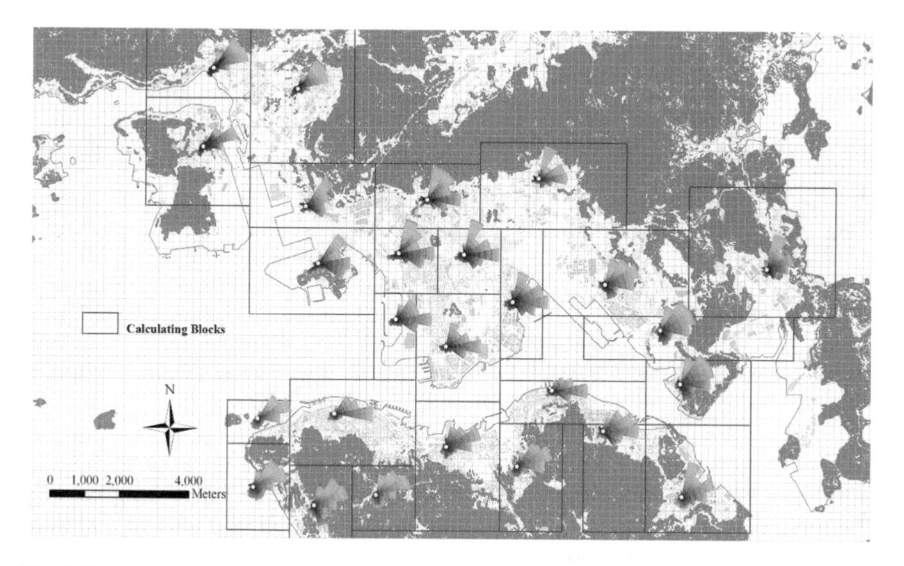

Fig. 2.6 Calculating blocks and representative wind roses at the height of 60 m in 16 directions for annual non-typhoon winds (01/01/2004–12/30/2004). *Data source* Institute for the Environment (IENV) (2010)

maintained by the Hong Kong Observatory, in 2004, the CALMET model adjusts the estimated meteorological fields for the kinematic effects of terrain, slope flows, and terrain blocking effects to reflect the impact of a fine-scale terrain on resultant wind fields at 100 m resolutions (Yim et al. 2007). In the CALMET model simulation, the vertical coordinates were set with 10 levels: 10, 30, 60, 120, 230, 450, 800, 1250, 1750, and 2600 m (Yim et al. 2007).

2.3.3 Calculation of $\lambda_{f(z)}$ in Grids with Uniform Size

In this chapter, $\lambda_{f(z)}$ was calculated in uniform grids. Each grid represents a local roughness value. The calculating boundary (grid boundary) was so small that large commercial podiums and public transport stations can be larger than the grid cell and cross the grid boundaries, as shown in Fig. 2.7. Values of $\lambda_{f(z)}$ for the cells at the middle of such large buildings may be underestimated. Therefore, to estimate the local roughness of every grid when buildings cross grid cells, a new method of which the cross-section areas (red areas) were included in the frontal areas of the corresponding grid cell was proposed. Compared with the map in polygon units (Gál and Unger 2009), this new calculation with uniform grid allows an exploration of mapping with a better explanatory power.

Compared with the conventional calculation of $\lambda_{f(z)}$, the nonexisting cross-section walls in the new calculation method could result in unrealistic surface roughness.

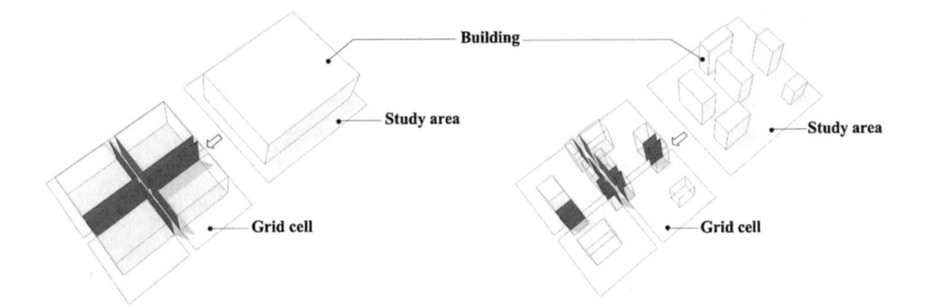

Fig. 2.7 Cross sections (red area) in two examples. When the study area is uniformly divided into four grid cells, the cross sections are generated

However, such cross sections may be needed to avoid underestimation of the surface roughness at the high-density urban areas covered by large and closely packed buildings. Thus, the correlations between the $VR_{\_w,j}$ and between the new method of $\lambda_{f(z)}$ with cross section and the traditional method of $\lambda_{f(z)}$ without cross section were compared.

Wind velocity ratios were obtained from wind tunnel tests for Hong Kong (HKPD 2008). The values of $VR_{\_w,j}$ of 10 study areas in wind tunnel tests were used as shown in Fig. 2.8. In the wind tunnel tests, test points were uniformly distributed in each study area. $VR_{\_w,j}$ for each test point has been described by (HKPD 2008):

$$VR_{\_w,j} = \sum_{i=1}^{16} P_i \cdot VR_{\_500,i,j}, \tag{2.4}$$

where P_i represents the annual probability of winds approaching the study area from the wind direction (i), and $VR_{\_500,i,j}$ represents the directional wind velocity ratio of the jth test point, the mean wind speed at 2 m above the ground with respect to the reference at 500 m (HKPD 2008). $VR_{\_500,i,j}$ is defined as (HKPD 2008):

$$VR_{\_500,i,j} = \frac{V_{\_p,i,j}}{V_{\_500,i}} \tag{2.5}$$

where $V_{\_p,i,j}$ represents the mean wind speed of the jth test point at the pedestrian level (2 m above the ground) for wind direction (i), and $V_{\_500,i}$ represents the mean wind speed of the jth test point at 500 m for wind direction (i).

As emphasized in Fig. 2.9, if study areas in wind tunnel experiment were crossed by grids in the map, the average of $\lambda_{f(z)}$ for the study areas is calculated by

$$\lambda_{f(z)} = \frac{\sum_{i=1}^{4} \lambda_{fi(z)} \cdot S_i}{S_t}, \tag{2.6}$$

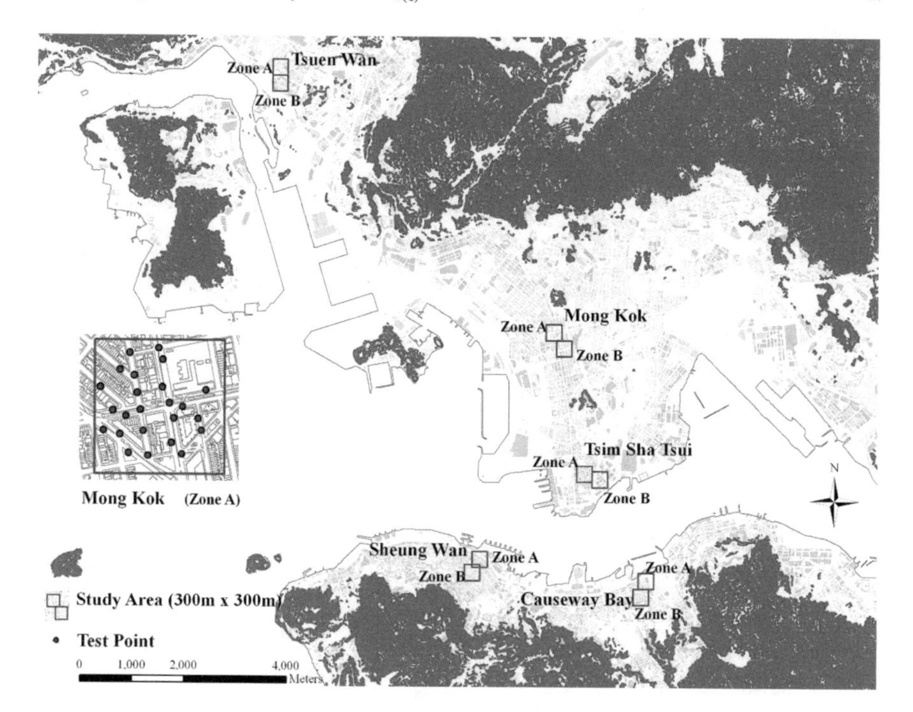

Fig. 2.8 Study areas in the wind tunnel experiment

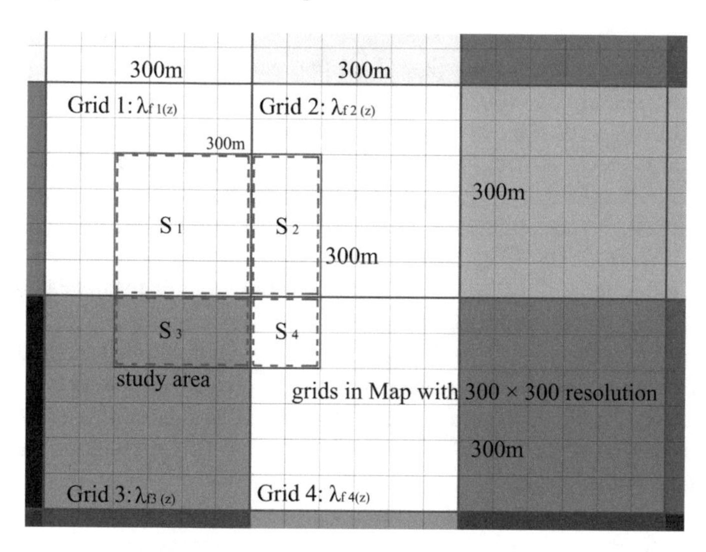

Fig. 2.9 Calculation of the average of $\lambda_{f(z)}$ in the study area (suppose the resolution is 300 × 300 m)

where $\lambda_{f\,i(z)}$ represents the frontal area density in the ith grid, S_i represents the area of the ith grid in the study area, and S_t represents the area of the study.

Table 2.1 Correlation between $VR_{\_w,j}$ and $\lambda_{f(0\text{–}15\ m)}$ in different resolutions and calculation methods

	R^2 (with cross sections)	R^2 (no cross sections)
Resolution: 300 × 300 m	0.96	0.96
Resolution: 200 × 200 m	0.87	0.88
Resolution: 100 × 100 m	0.71	0.70
Resolution: 50 × 50 m	0.63	0.66

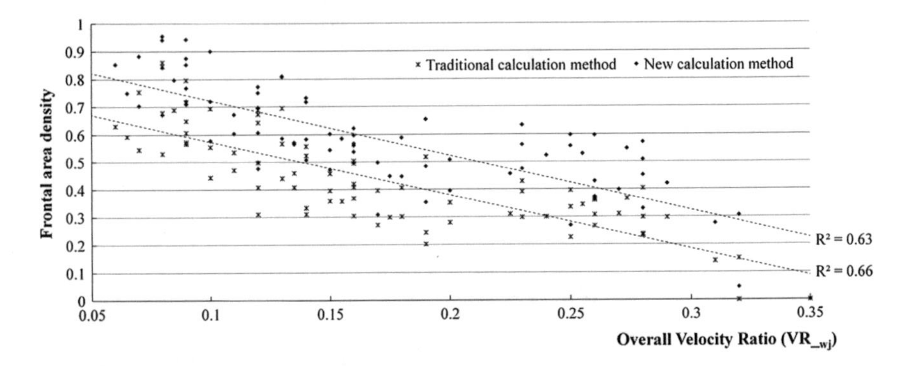

Fig. 2.10 Relationships between $VR_{\_w,j}$ and $\lambda_{f(0\text{–}15\ m)}$ calculated by different methods in 50 × 50 m resolutions. The number of the point pairs is 80, and the significance level is 5%

The $\lambda_{f(z)}$ in the podium layer ($\lambda_{f(0\text{–}15\ m)}$) that corresponds to the four grid sizes (resolutions), namely 50, 100, 200, and 300 m, were calculated. The R^2 values in Table 2.1 illustrate that the new calculation method can be as accurately predict the wind velocity ratio as the traditional method without cross section. As expected, in accordance with values of the $\lambda_{f(0\text{–}15\ m)}$ including the unreal flow-confronting areas were larger than the ones calculated by the traditional method, and their correlations with $VR_{\_w,j}$ were similar (Fig. 2.10).

On the other hand, the $\lambda_{f(0\text{–}15\ m)}$ values in the Kowloon Peninsula calculated by the two methods were compared. In high-density urban areas with large and closely packed buildings, the $\lambda_{f(0\text{–}15\ m)}$ values calculated by the traditional method without cross section were less than 0.1; some of them were even close to 0. This is a serious underestimation to the surface roughness. Highlighted in Fig. 2.11, the new method with cross section in this chapter efficiently alleviates these underestimations by including the cross sections.

Based on the regression analysis result, following understandings can be stated: the new calculation method with cross sections can correctly predict the wind velocity ratio. Furthermore, compared with the traditional method of calculating frontal area density, the new method can alleviate the underestimation of mapping urban surface roughness in high-density cities with large and closely packed buildings.

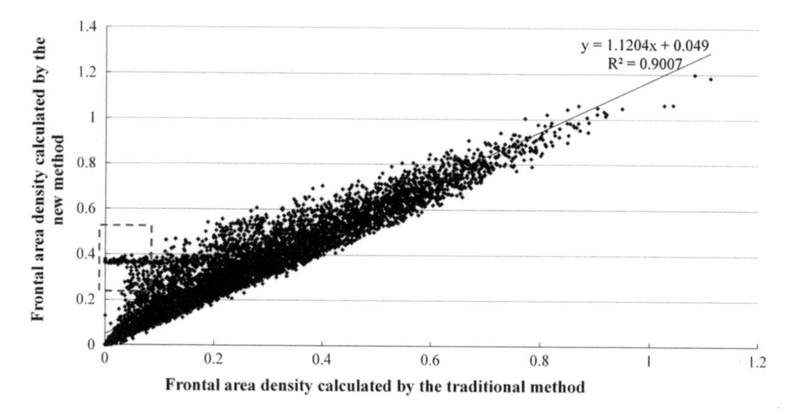

Fig. 2.11 Cross-comparison between $\lambda_{f(0\text{–}15\ m)}$ calculated by different methods in 50 × 50 m resolutions (Test area: Kowloon Peninsula). The number of the point pairs is 7519, and the significance level is 5%

2.3.4 *Grid Sensitivity (Resolution)*

As shown in Table 2.1, the values of R^2 decrease along with the reduction of the grid sizes. Choosing a larger grid size would have a positive effect on depicting the urban wind environment. However, R^2 should not be the only criterion for selecting one grid size over another. For mapping roughness, the explanatory power of the map should not be totally traded off for the sake of the correctness of $\lambda_{f(z)}$. After weighing the considerations, the resolution of 200 × 200 m was adopted in mapping urban permeability in Hong Kong.

2.4 Development of Empirical Model

The skimming flow regime is normally found at the top of compact high-rise building areas (Letzel et al. 2008). Similarly, due to the signature urban morphology of Hong Kong (i.e., high-density and tall buildings), the airflow above the top of the urban canopy layer may not easily enter the deep street canyons to benefit the wind environment at the pedestrian level. Thus, the wind velocity ratio at the pedestrian level mostly depends on the wind permeability of the podium layer. A statistical study was conducted to validate the above assumption by comparing the sensitivities of $VR_{\_w,j}$ to changes of $\lambda_{f(z)}$ calculated at different layers. The cross-comparison results are plotted in Fig. 2.12a, b, which indicate that $VR_{\_w,j}$ has a higher correlation with $\lambda_{f(z)}$ at the podium layer (0–15 m). This illustrates that pedestrian-level wind speed depends on the urban morphology at the podium layer (0–15 m), rather than the building layer (15–60 m) or the whole canopy layer (0–60 m).

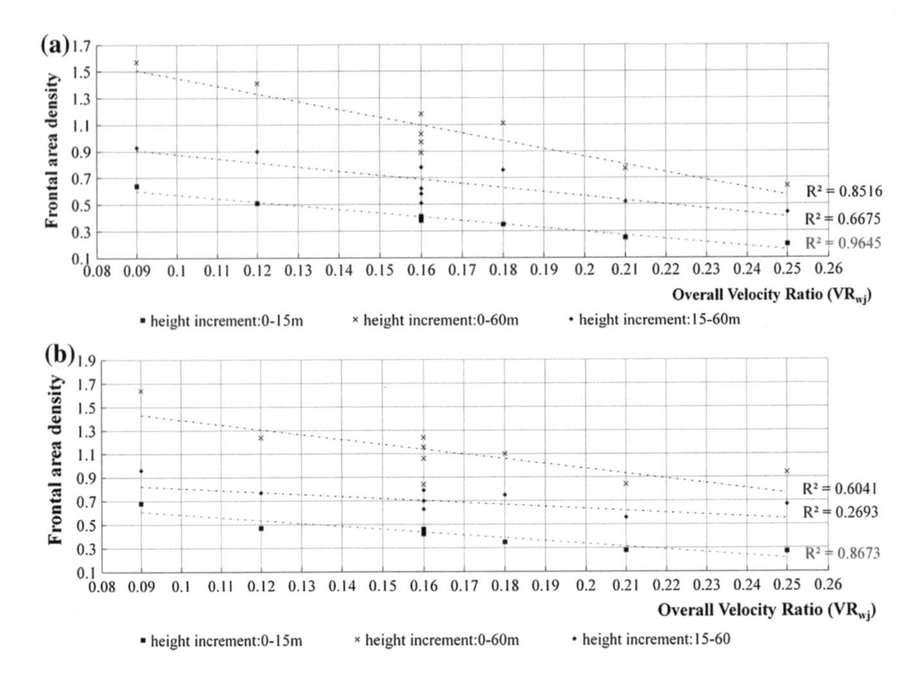

Fig. 2.12 **a** Relationships between overall velocity ratio ($VR_{\_w,j}$) and averaged $\lambda_{f(0\text{–}15\ m)}$, $\lambda_{f(15\text{–}60\ m)}$, and $\lambda_{f(0\text{–}60\ m)}$ in 300 × 300 m resolutions. The number of the point pairs is 9, and the significance level is 5%. **b** Relationships between overall velocity ratio ($VR_{\_w,j}$) and averaged $\lambda_{f(0\text{–}15\ m)}$, $\lambda_{f(15\text{–}60\ m)}$, and $\lambda_{f(0\text{–}60\ m)}$ in 200 × 200 m resolutions. The number of the point pairs is 9, and the significance level is 5%

2.5 Implementation in Urban Planning

2.5.1 Mapping Urban Wind Permeability Using $\lambda_{f(z)}$

This understanding is important to support the evidence-based urban planning and design in order to improve the pedestrian-level wind environment at high-density urban areas. Compared with front area index, which was used to detect the air paths in Hong Kong (Wong et al. 2010), $\lambda_{f(0\text{–}15\ m)}$ has been proven to be a better morphological factor in depicting the wind environment at the pedestrian level. As shown in Fig. 2.13a, the map of the frontal area density (0–15 m) depicts the local wind permeability at the podium layer in the Kowloon Peninsula and Hong Kong Island. The continuous belts of high surface roughness on the northern coastline of the Hong Kong Island and both sides of the Kowloon peninsula, referred to as the wall effect of the Kowloon Peninsula, are evident (Yim et al. 2009), whereas the wind permeability is very low. The maps of the frontal area density (0–60 m and 15–60 m) are also presented, as shown in Fig. 2.13b, c. These two maps are important for describing the wind permeability at the urban canyon layer. The turbulent mixing at the urban canyon layer is essential to improve urban air ventilation, alleviate air pollution, and dissipate the anthropogenic heat.

2.5.2 Ground Coverage Ratio and Frontal Area Density

Compared with $\lambda_{f(z)}$, ground coverage ratio (GCR) is a two-dimensional parameter commonly used by urban planners. GCR is defined as

$$\text{GCR} = \frac{A_b}{A_T} = \frac{w^2 \cdot n}{A_T}, \; (n \geq 1) \, , \tag{2.7}$$

where A_T represents the site area, A_b represents the built area, w represents the average building width, and n represents the number of buildings. A statistical study was conducted to convert the frontal area density analysis to a practical design and planning tool; this was accomplished by investigating the relationship between $\lambda_{f(0\text{–}15\,m)}$ and GCR.

Local values of $\lambda_{f(0\text{–}15\,m)}$ and the GCR of the 1004 test areas (200 × 200 m) in the Kowloon Peninsula and the Hong Kong Island were calculated. Figure 2.14 shows a good linear relationship of both ($R^2 = 0.77$). However, it should be noted the presence of outlier values of local surface roughness of large podiums and industrial buildings.

Equation 2.8 indicates the relationship between GCR and $\lambda_{f(z)}$, which depends on k, the ratio between the averaged building width (w) and podium layer height, i.e. 15 m. If the building width of urban areas is much larger than that of other areas with normal building morphology, the correlation between GCR and $\lambda_{f(0\text{–}15\,m)}$ in such areas can be significantly different from other areas. Four examples of such sites, points A–D, are shown in Figs. 2.14 and 2.15.

$$\text{GCR} = \frac{w \cdot \lambda_{f(0-15\,m)}}{15} = k \cdot \lambda_{f(0-15\,m)}, \quad k = \frac{w}{15}. \tag{2.8}$$

Based on the above discussion, following understandings can be stated:

(1) There is a good linear relationship between $\lambda_{f(0\text{–}15\,m)}$ and GCR ($R^2 = 0.77$) in most of the test points. For planners, using GCR to predict the wind environment at the pedestrian level is reasonable. Compared with other traditional maps (Gál and Unger 2009; Wong et al. 2010), the proposed GCR map is more applicable to urban designers and planners due to its accessibility to the planners in the planning process.
(2) Local values of some areas may deviate due to the extremely large building widths (large commercial podiums and industrial buildings). In this type of areas, the wind permeability cannot be predicted in GCR. However, the occurrence of this type of extreme examples is very small (approximately 2%).

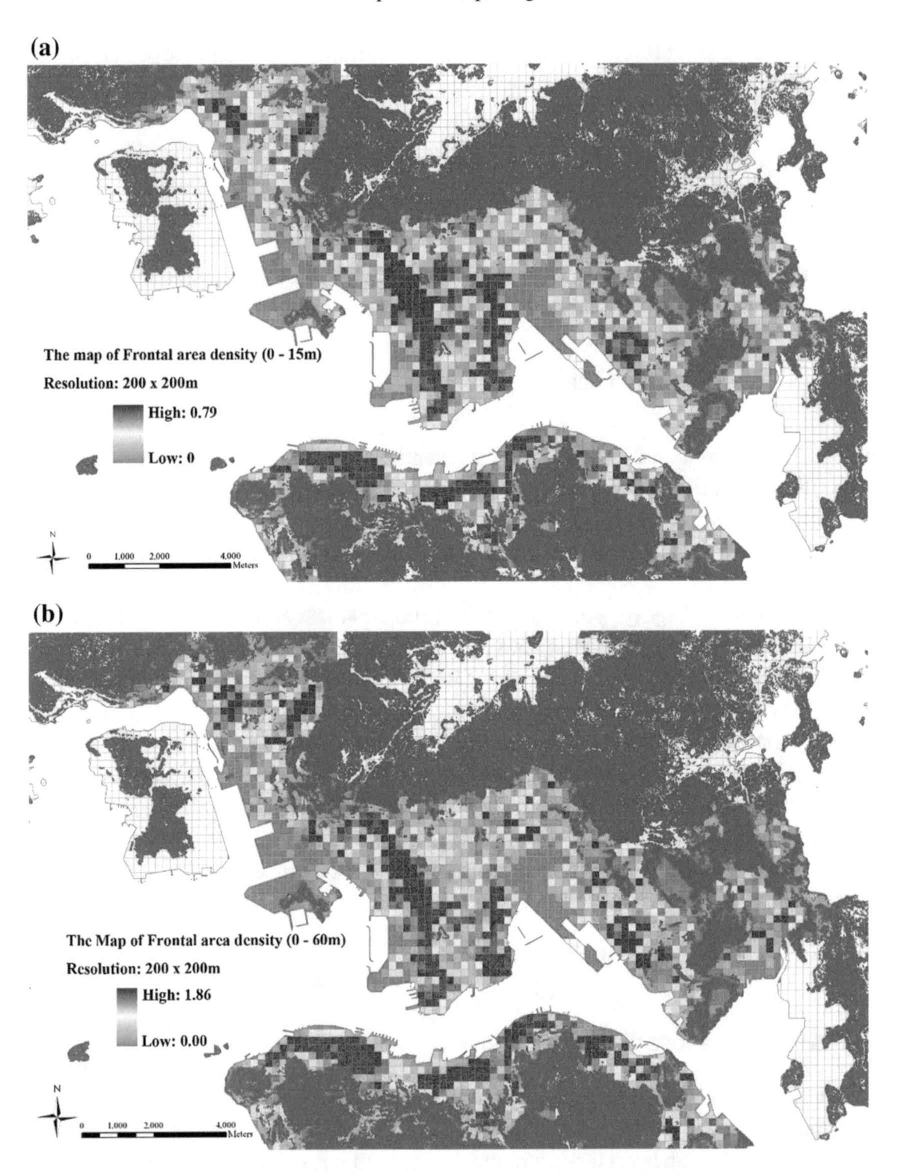

Fig. 2.13 **a** Map of frontal area density in Kowloon and Hong Kong Island (Resolution: 200 × 200 m, height increment (Δz): 0–15 m) (For interpretation of the references to color in the text, the reader is referred to the electronic version of this book). **b** Map of frontal area density in Kowloon and Hong Kong Island (Resolution: 200 × 200 m, height increment (Δz): 0–60 m) (For interpretation of the references to color in the text, the reader is referred to the electronic version of this book). **c** Map of frontal area density in Kowloon and Hong Kong Island (Resolution: 200 × 200 m, height increment (Δz): 15–60 m (For interpretation of the references to color in the text, the reader is referred to the electronic version of this book)

(c)

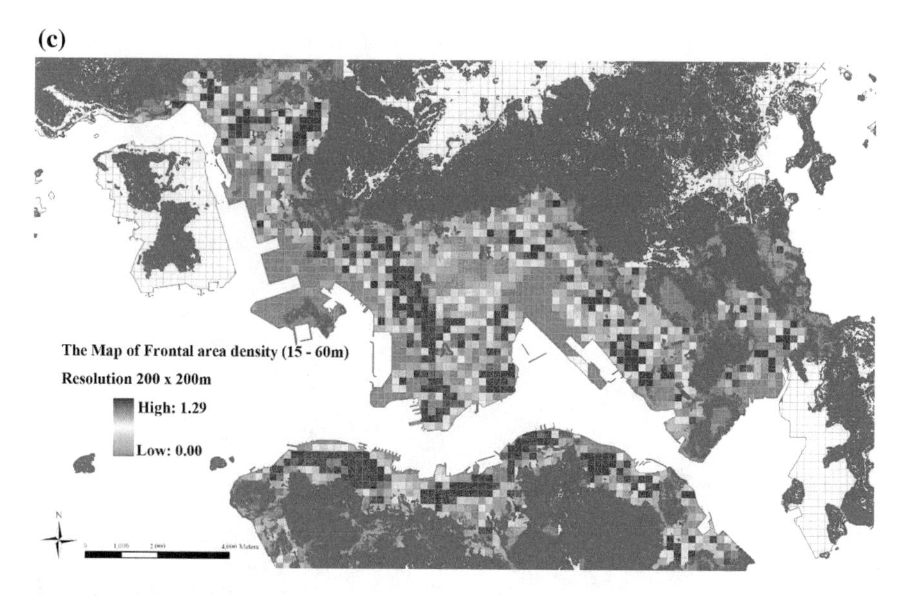

Fig. 2.13 (continued)

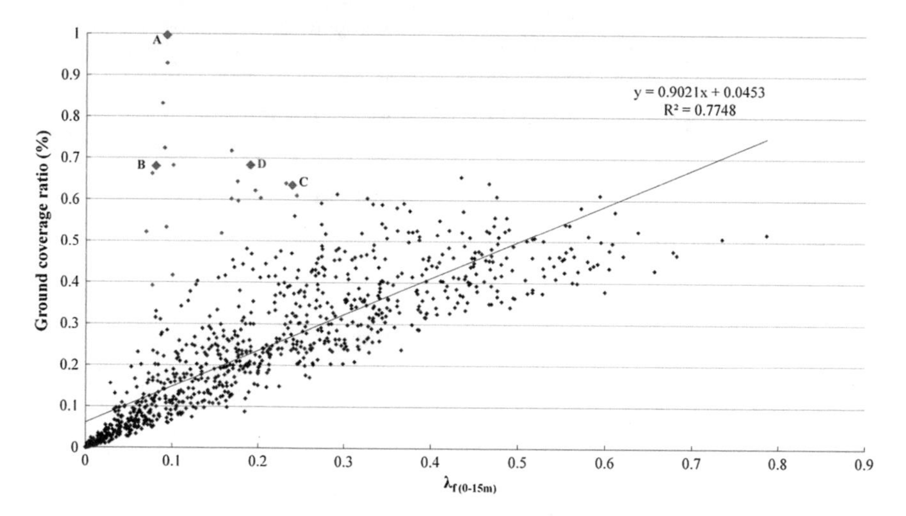

Fig. 2.14 Relationship between $\lambda_{f(0\text{–}15\ m)}$ and GCR in Kowloon Peninsula and Hong Kong Island. The outlier values were highlighted in red. The points A, B, C, and D correspond to the four examples in Fig. 2.15. The number of the point pairs is 1032, and the significance level is 5%

2.5.3 Mapping Urban Wind Permeability Using GCR

An urban-level wind environment of Hong Kong was mapped using GCR information in this section. Kubota et al. (2008) and Yoshie et al. (2008) conducted an earlier investigation on the relationship between GCR and the spatial average of wind veloc-

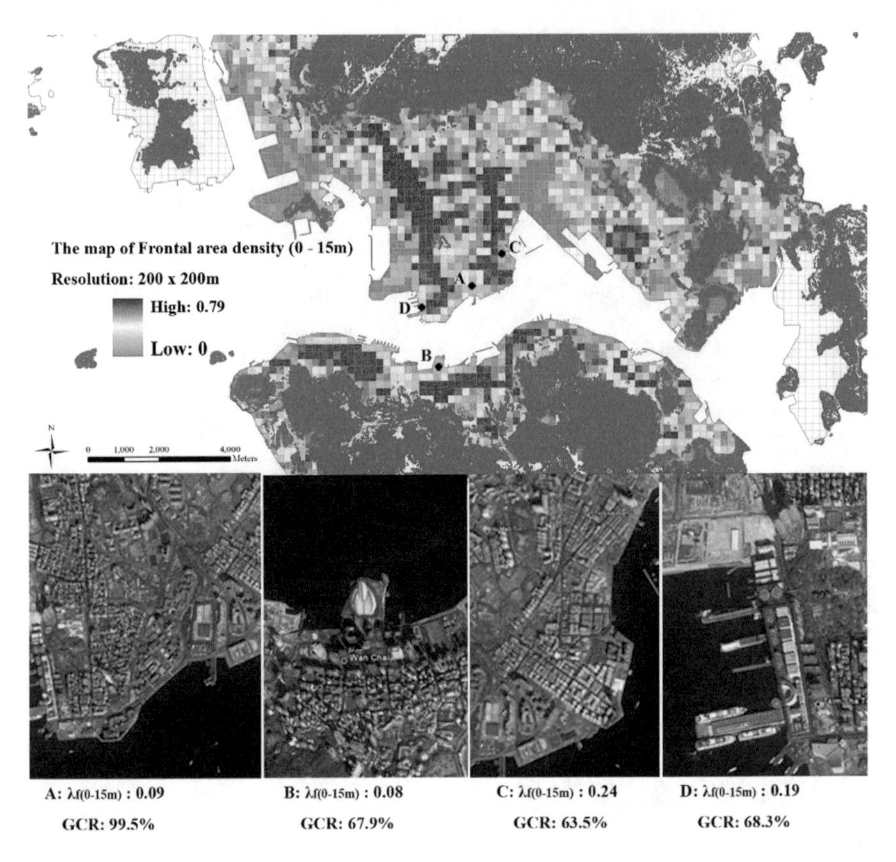

Fig. 2.15 Four typical cases highlighted in Fig. 2.14

ity ratios at a height of 1.5 m obtained by wind tunnel tests, both in Japanese cities and in the Mong Kok area of Hong Kong. This relationship can be used as the basis for the threshold values of the map classification. Coupled with the classification, the effect of different GCRs on the wind permeability can be identified. As shown in Fig. 2.16, three classification values are assigned: "Class 1", "Class 2", and "Class 3", which denote good, reasonable, and poor pedestrian wind performance, respectively.

Based on this classification, the map of wind performance at the podium layer in Hong Kong was generated as shown in Fig. 2.17. Compared with the roughness map without classification, the map in this study is more intuitive; in addition, it allows urban planners to better modify building morphology in order to improve the urban air environment. Such map can be the spatial reference for urban planners.

After incorporating the respective site-specific wind roses, the areas with low wind permeability are depicted in Fig. 2.17. These areas block wind and worsen the wind environment at the pedestrian level of their leeward districts. Potential air paths in the podium layer are also marked out in this map. The potential air paths would play an important role to improve the urban ventilation and environment quality by

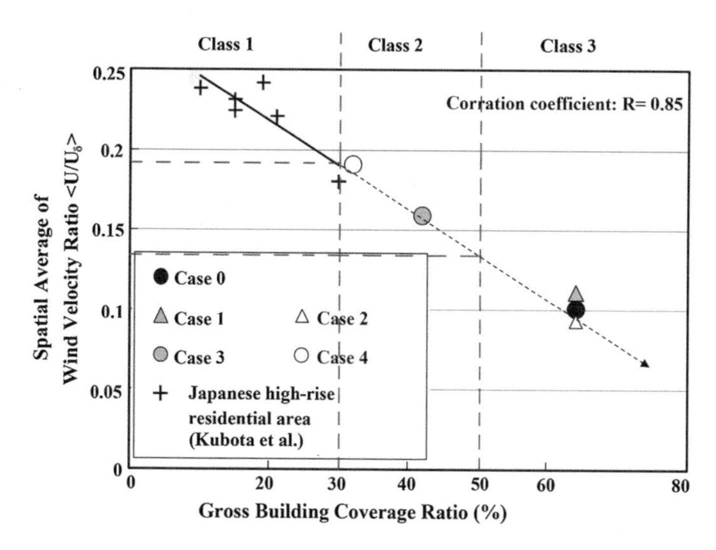

Fig. 2.16 Relationship between ground coverage ratio and spatial average of wind velocity ratios in Mong Kok and cities in Japan (Kubota et al. 2008; Yoshie et al. 2008, edited by authors). The number of the point pairs is 11

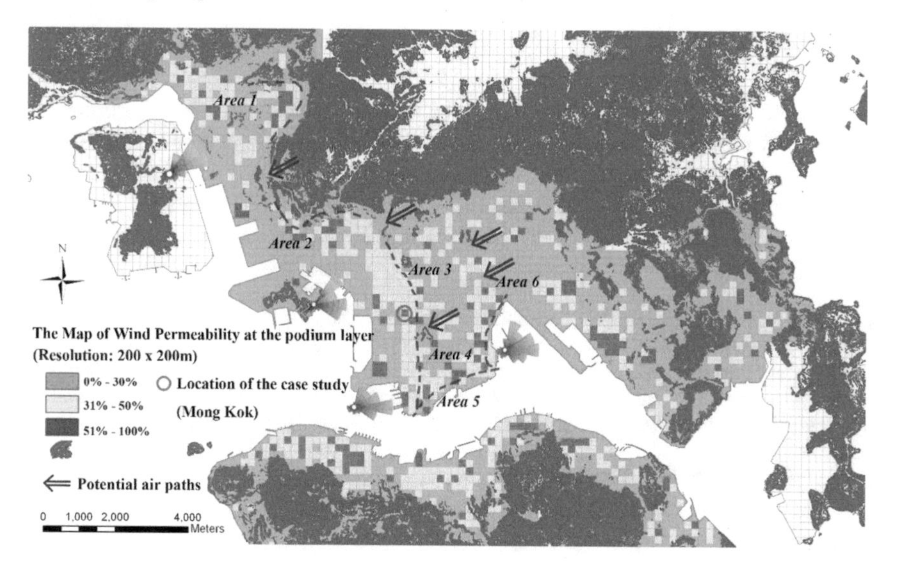

Fig. 2.17 Map of wind permeability at the podium layer. The wind permeability at the podium level is depicted in the map: Class 1: GCR = 0–30%, Class 2: GCR = 31–50%, and Class 3: GCR > 50%. Based on the respective annual prevailing wind direction, the areas with low wind permeability are pointed out. These areas could block the natural ventilation and worsen the leeward districts' wind environment at the pedestrian level. Potential air paths in the podium layer are also marked out in this map (For interpretation of the references to color in the text, the reader is referred to the electronic version of this book)

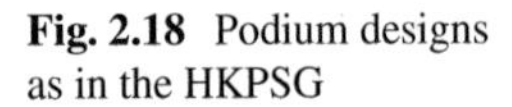

Fig. 2.18 Podium designs as in the HKPSG

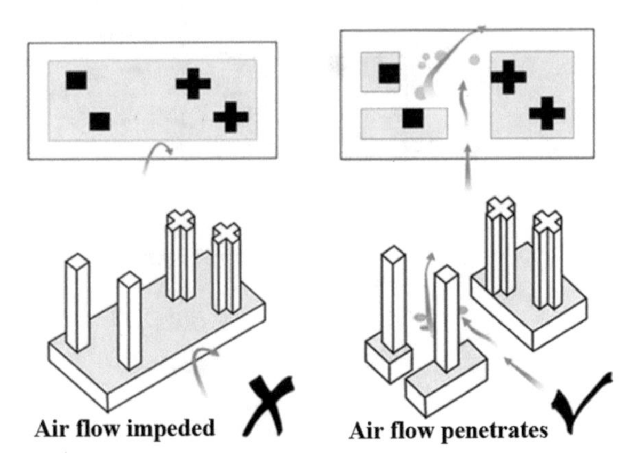

bringing fresh airflow into the urban areas for the purpose of dissipating air pollutant and mitigating urban heat island intensity.

2.6 Conclusions

The chapter has highlighted a number of important points that should be considered by city planners. First, one of the most significant factors is urban morphology, especially the podium layer, and its implication to the urban air ventilation environment. According to Chap. 11, Sects. 9–13 of the Hong Kong Planning Standards and Guidelines (HKPSG) (HKPD 2006), a number of urban forms deemed to be conducive to the urban air ventilation environment have been proposed:

> … it is critical to increase the permeability of the urban fabric at the street levels. Compact integrated developments and podium structures with full or large ground coverage on extensive sites typically found in Hong Kong are particularly impeding air movement and should be avoided where practicable. The following measures should be applied at the street level for large development/redevelopment sites particularly in the existing urban areas:
>
> - providing setback parallel to the prevailing wind;
> - designating non-building areas for sub-division of large land parcels;
> - creating voids in facades facing wind direction; and/or
> - reducing site coverage of the podia to allow more open space at grade.

Where appropriate, a terraced podium design should be adopted to direct downward airflow to the pedestrian level as shown in Figs. 2.18 and 2.19.

This chapter shows that the qualitative understanding of the podium structure, as mentioned in the HKPSG, is valid. In Hong Kong, some areas of high podium coverage can be identified. These areas require the most significant design, planning intervention and improvement.

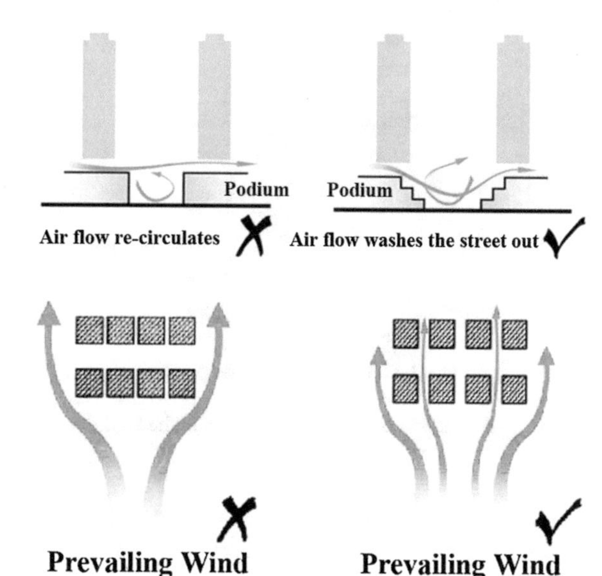

Fig. 2.19 Terraced podium designs as in the HKPSG

Fig. 2.20 Building dispositions as in the HKPSG

For building block disposition, the chapter has emphasized that city planners need to factor in the prevailing wind understanding to street layout and building disposition design as shown in Fig. 2.20. This understanding is in line with the concerns of the so-called "wall buildings", wherein a line of tall buildings screen the waterfront from the inland areas, thereby blocking the incoming urban air ventilation from the sea.

Based on the GCR information readily available to planners working on their GIS system, the chapter has shown that planners can easily generate an urban wind permeability map, thereby enhancing the possibility to identify problem areas and, more importantly, to emphasize on potential air paths. The map also enables the interconnectivity of open spaces for urban air ventilation, and allows planners to take urban breezeways into account and design in accordance with the recommendations of the HKPSG (HKPD 2006), as shown in Fig. 2.21:

> For better urban air ventilation in a dense, hot-humid city, breezeways along major prevailing wind directions and air paths intersecting the breezeways should be provided in order to allow effective air movements into the urban area to remove heat, gases and particulates and to improve the micro-climate of urban environment.
>
> Breezeways should be created in forms of major open ways, such as principal roads, inter-linked open spaces, amenity areas, non-building areas, building setbacks and low-rise building corridors, through the high-density/high-rise urban form. They should be aligned primarily along the prevailing wind direction routes, and as far as possible, to also preserve and funnel other natural airflows including sea and land breezes and valley winds, to the developed area.
>
> The disposition of amenity areas, building setbacks and non-building areas should be linked, and widening of the minor roads connecting to major roads should be planned in such a way to form ventilation corridors/air paths to further enhance wind penetration into inner parts of urbanized areas. For effective air dispersal, breezeways and air paths should be perpendicular or at an angle to each other and extend over a sufficiently long distance for continuity.

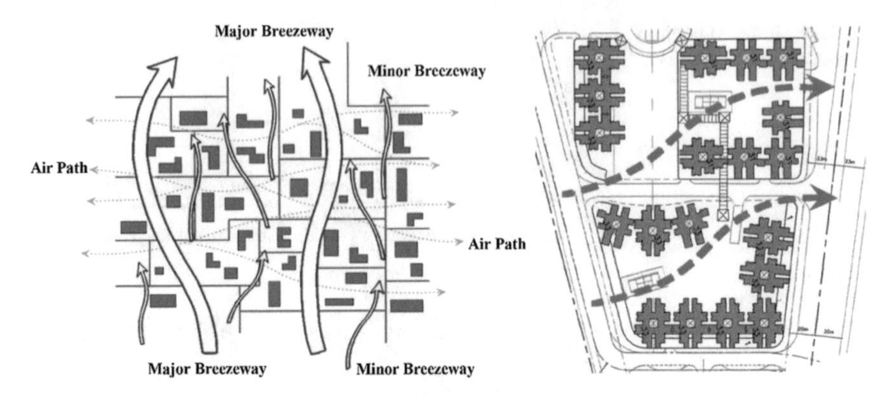

Fig. 2.21 Breezeway and air path design as in the HKPSG

Using the urban wind permeability map of the territory, city planners can initially estimate the possible urban air ventilation environment of the urban areas with the average velocity ratios. Adjusting the pedestrian-level wind speeds and predicting the bioclimatic conditions of the city have become possible.

Overall, the chapter has demonstrated a practical and reliable way for city planners to quickly obtain district-level urban air ventilation information for their board-based design works at the early stages. Conceptually, avoiding wrong decisions that may be difficult to rectify later is, therefore, possible.

References

Architectural Institute of Japan (AIJ) (2007) AIJ guidebook for practical applications of CFD to pedestrian wind environment around buildings (ISBN 978-4-8189-2665-3)

Arnfield AJ (2003) Two decades of urban climate research: a review of turbulence, exchanges of energy and water, and the urban heat island. Int J Climatol 23(1):1–26

Ashie Y, Hirano K, Kono T (2009) Effects of sea breeze on thermal environment as a measure against Tokyo's urban heat island. Paper presented at the seventh international conference on urban climate, Yokohama, Japan

Bottema M (1996) Roughness parameters over regular rough surfaces: Experimental requirements and model validation. J Wind Eng Ind Aerodyn 64(2–3):249–265

Burian SJ, Velugubantla SP, Brown MJ (2002) Morphological analyses using 3D building databases: Phoenix, Arizona. Los Alamos National Laboratory, LA-UR-02-6726

Davenport AG, Grimmond CSB, Oke TR, Wieringa J (2000) Estimating the roughness of cities and sheltered country. Paper presented at the proceedings of the 12th conference on applied climatology, Boston

Dudhia J (1993) A nonhydrostatic version of the penn state-NCAR mesoscale model: validation tests and simulation of an Atlantic cyclone and cold front. Mon Weather Rev 121:1493–1513

Frank J (2006) Recommendations of the COST action C14 on the use of CFD in predicting pedestrian wind environment. J Wind Eng 108:529–532

Gál T, Unger J (2009) Detection of ventilation paths using high-resolution roughness parameter mapping in a large urban area. Build Environ 44(1):198–206

Grimmond CSB, Oke TR (1999) Aerodynamic properties of urban areas derived from analysis of surface form. J Appl Meteorol 38:1262–1292
Hong Kong Planning Department (HKPD) (2005) Feasibility study for establishment of air ventilation assessment system. Final report, The government of the Hong Kong Special Administrative Region
Hong Kong Planning Department (HKPD) (2006) Hong Kong planning standards and guidelines. The government of the Hong Kong Special Administrative Region
Hong Kong Planning Department (HKPD) (2008) Urban climatic map and standards for wind environment—feasibility study. Working paper 2B: wind tunnel benchmarking studies, Batch I. The government of the Hong Kong Special Administrative Region
Hong Kong Rating and Valuation Department (HKRVD) (2009) Private office—1984–2007 Rental and price indices for Grade A Office in core districts. The government of the Hong Kong Special Administrative Region
Institute for the Environment (IENV) (2010) Study of ventilation over Hong Kong. The Hong Kong University of Science and Technology
Kastner-Klein P, Fedorovich E, Rotach MW (2001) A wind tunnel study of organised and turbulent air motions in urban street canyons. J Wind Eng Ind Aerodyn 89(9):849–861
Kondo H, Asahi K, Tomizuka T, Suzuki M (2006) Numerical analysis of diffusion around a suspended expressway by a multi-scale CFD model. Atmos Environ 40(16):2852–2859
Kubota T, Miura M, Tominaga Y, Mochida A (2008) Wind tunnel tests on the relationship between building density and pedestrian-level wind velocity: development of guidelines for realizing acceptable wind environment in residential neighborhoods. Build Environ 43(10):1699–1708
Kutzbach J (1961) Investigations of the modifications of wind profiles by artificially controlled surface roughness. University of Wisconsin—Madison, Madison
Landsberg HE (1981) The urban climate, vol 28. Academic Press, INC. (London) LTD., London
Lettau H (1969) Note on aerodynamic roughness-parameter estimation on the basis of roughness-element description. J Appl Meteorol 8:828–832
Letzel MO, Krane M, Raasch S (2008) High resolution urban large-eddy simulation studies from street canyon to neighbourhood scale. Atmos Environ 42(38):8770–8784
MacDonald RW, Griffiths RF, Hall DJ (1998) An improved method for the estimation of surface roughness of obstacle arrays. Atmos Environ 32(11):1857–1864
Mochida A, Murakami S, Ojima T, Kim S, Ooka R, Sugiyama H (1997) CFD analysis of mesoscale climate in the Greater Tokyo area. J Wind Eng Ind Aerodyn 67–68:459–477
Murakami S, Ooka R, Mochida A, Yoshida S, Kim S (1999) CFD analysis of wind climate from human scale to urban scale. J Wind Eng Ind Aerodyn 81(1–3):57–81
Ng E (2007) Feasibility study for establishment of air ventilation assessment system (AVAS). HKIP J 22(1):39–45
Ng E (2009) Policies and technical guidelines for urban planning of high-density cities—air ventilation assessment (AVA) of Hong Kong. Build Environ 44(7):1478–1488
Oke TR (1987) Boundary layer climates, 2nd edn. Methuen Inc, USA
Perry SG, Heist DK, Thompson RS, Snyder WH, Lawson RE (2004) Wind tunnel simulation of flow and pollutant dispersal around the World Trade Centre site. Environ Manager 31–34
Plate EJ (1999) Methods of investigating urban wind fields—physical models. Atmos Environ 33(24–25):3981–3989
Ranade MB, Woods MC, Chen FL, Purdue LJ, Rehme KA (1990) Wind tunnel evaluation of PM10 samplers. Aerosol Sci Technol 13(1):54–71
Ratti C, Sabatino SD, Britter R, Brown M, Caton F, Burian S (2002) Analysis of 3-D urban databases with respect to pollution dispersion for a number of European and American cities. Water Air Soil Pollut Focus 2:459–469
Raupach MR (1992) Drag and drag partition on rough surfaces. Bound-Layer Meteorol 60:375–395
Scire JS, Robe FR, Fernau ME, Yamartino RJ (2000) A user's guide for the CALMET meteorological model (Version 5). Earth Tech Inc, Concord, MA, USA

Shao Y, Yang Y (2005) A scheme for drag partition over rough surfaces. Atmos Environ 39(38):7351–7361

Tominaga Y, Mochida A, Yoshie R, Kataoka H, Nozu T, Yoshikawa M, Shirasawa T (2008) AIJ guidelines for practical applications of CFD to pedestrian wind environment around buildings. J Wind Eng Ind Aerodyn 96:1749–1761

Williams CD, Wardlaw RL (1992) Determination of the pedestrian wind environment in the city of Ottawa using wind tunnel and field measurements. J Wind Eng Ind Aerodyn 41(1–3):255–266

Wong MS, Nichol JE, To PH, Wang J (2010) A simple method for designation of urban ventilation corridors and its application to urban heat island analysis. Build Environ 45(8):1880–1889

Yim SHL, Fung JCH, Lau AKH, Kot SC (2007) Developing a high-resolution wind map for a complex terrain with a coupled MM5/CALMET system. J Geophys Res 112:D05106

Yim SHL, Fung JCH, Lau AKH, Kot SC (2009) Air ventilation impacts of the "wall effect" resulting from the alignment of high-rise buildings. Atmos Environ 43(32):4982–4994

Yoshie R, Tanaka H, Shirasawa T, Kobayashi T (2008) Experimental study on air ventilation in a built-up area with closely-packed high-rise building. J Environ Eng 627:661–667(in Japanese)

Chapter 3
Implementation of Morphological Method in Urban Planning

3.1 Introduction

Urban environment has been changed and deteriorated, due to rapid urbanization, as well as other factors including a lack of implementation of environmental information and knowledge in the urban planning practice. Therefore, there is a need to develop a systematic and user-friendly method for city planners and policymakers to make scientific and evidence-based decisions in order to address urban environmental issues.

Germany and Japan have integrated pioneering work in the field of urban climatic application into urban planning (Ren et al. 2011). German cities have protected their urban environment carefully in the local development since the 1950s. Stuttgart Municipal Government has been making a continuous effort to upgrade the air quality of Stuttgart. One of their useful measures is the air path development scientists, urban planners, and local governors worked together to evaluate the air-flow distribution patterns, detect the possible air paths that can bring the fresh air from the surrounding hillsides to the downtown areas of Stuttgart, and control the urban development carefully and strategically (Baumueller et al. 2009). Relevant plan actions have also played an important role in mitigating urban heat island and improving air quality. Correspondingly, the Japanese Government and their researchers had focused their attention on the wind environment since the 1990s. The Tokyo Metropolitan Government that oversees eight main counties finished a study on air path in 2007 to provide a collection of relevant wind information (e.g., wind rose, annual and seasonal prevailing wind information, and land–sea breezes system), as well as detailed plan of

---

Originally published in Chao Yuan, Chao Ren and Edward Ng, 2014. GIS-based surface roughness evaluation in the urban planning system to improve the wind environment–A study in Wuhan, China. Urban Climate 10, pp. 585–593,

C. Yuan, *Urban Wind Environment*, SpringerBriefs in Architectural Design and Technology, https://doi.org/10.1007/978-981-10-5451-8_3

developing air paths in Tokyo Metropolitan areas for their planners (Architectural Institute of Japan (AIJ) 2008).

Hong Kong is one of the densest and most populated cities in the world. Natural ventilation in urban planning is a big challenge to local planners and governors, as the compact building blocks in the city create a "wall-effect", which highly interferes with local air circulation. Local planners and governors have worked together in recent years to develop a wind information layer for planning use based on the available meteorological records, CFD simulations and expert evaluation (Ng 2012); and this layer has been used by the Planning Department of Hong Kong Government to guide the new town plan and urban renewal.

3.2 Objectives

The study of outdoor natural ventilation often requires large-scale aerodynamics modeling. Both physical modelling (wind tunnel) and numerical modeling can provide data regarding the airflow within the urban canopy layer. However, conducting these modeling tests for a particular urban planning exercise is expensive and time consuming. Since modeling results obtained from the traditional modeling methods cannot keep up with the quick planning processes, a novel methodology (see Chap. 2) that uses a surface roughness understanding of the urban morphological implication to the urban wind environment is a more useful alternative for urban planners. Given the growing concerns related to how urban environment is evaluated and how air paths are detected in order to meet the requirements of practical urban planning, this chapter aims to:

- Analyze urban permeability to detect potential air paths in order to improve urban performance in outdoor natural ventilation;
- Highlight the implementation of modeling results in urban planning practices, and interweave the modeling results into different urban planning stages and scales, such as the master and district planning.

Specifically, this chapter takes the city of Wu Han, China as an example to illustrate how to apply the new morphological model (introduced in Chap. 2) into actual planning practice. Wuhan is one of the megacities in China that has undergone rapid urbanization in the past two decades. With an area of 8500 km^2 and a population of over 10 million, Wuhan is located inland and west of Shanghai. In the summer months, weather in Wuhan is relatively hot, with an average daytime temperature of 33 °C. For years, the city planners in Wuhan have been postulating the idea of urban air paths for the making of their master plan. This chapter demonstrates an application of the morphological modeling method with local GIS data (3D building database), which will enable the local planners to easily evaluate the urban permeability in order to understand the outdoor natural ventilation performance of the city and make evidence-based decision regarding urban planning.

3.3 Approach

3.3.1 Morphological Method

The basic assumption in the current morphological models, such as the models provided by MacDonald et al. (1998), Lettau (1969), and Bottema (1996), was stated by MacDonald et al. (1998) as: "*...we assume that there is negligible wake interference between the surface obstacles and that the mean velocity profile approaching each obstacle is logarithmic.*" Due to this assumption, the reason why these models are only valid when frontal area density (λ_f) is less than about 0.3–0.5 is obvious: the mean velocity profile approaching each obstacle is not logarithmic as the surface roughness increases, because the interference among obstacles promotes the recirculating flow that dominates the flow near the ground.

As a result, the displacement height (d) needs to be incorporated into the logarithmic velocity profile (MacDonald et al. 1998). If the wind speed is above the displacement height, the mean wind profile approaching the obstacle will become logarithmic again. Therefore, MacDonald et al. (1998) pointed out that frontal area density above the displacement height (λ_f^*) could be better to estimate z_0 than λ_f, developed an in-canopy logarithmic profile, and assumed the mean wind speed below the displacement height (a new smooth surface) to be zero in the derivation. The above analyses bring a great trouble to the practical application of the near ground wind speed estimation at high-density urban areas, as many reliable wind tunnel experiments have shown the wind speed below the displacement to be nonzero at high-density areas. Rather than the logarithmic profile model developed by MacDonald et al. (1998), the exponential solution was introduced by Cionco (1965) and Coceal and Belcher (2004) developed the in-canopy model by parameterizing the canopy element drag ($D_i\ (x, y, z)$) and turbulent mixing $\left\langle \overline{u_i' u_j'} \right\rangle$. But as Coceal and Belcher (2004) mentioned, a balance between sectional drag and shear stress is assumed in the exponential models, and the model for the spatially averaged mixing length (l_m) may fail when recirculating flow dominate the flow near the ground with very densely packed canopy element. We shall find later that λ_f in this chapter could be larger than 1.0 at metropolitan areas.

Therefore, the spatially averaged wind speed below the displacement height could depend on the building geometries such as λ_f' (frontal area density below the displacement height) in high-density areas, instead of λ_f^* by which the mean wind profile above the displacement height can be well identified. The cross-comparison conducted in Chap. 2, that is, the VR (the ratio of pedestrian-level wind speed to the wind speed at the reference height) is well related with $\lambda_f'(\lambda_{f(0-15\,m)})$, rather than $\lambda_f^*(\lambda_{f(15\text{–}60\,m)})$ and $\lambda_{f(0\text{–}60\,m)}$, supported the above hypothesis.

By using a high-resolution (1 × 1 m) building height database, a self-developed program embedded as a VBA script in the ArcGIS system is applied to calculate the frontal area density $\lambda_{f(z)}$. The specific settings for Wuhan city as the background are

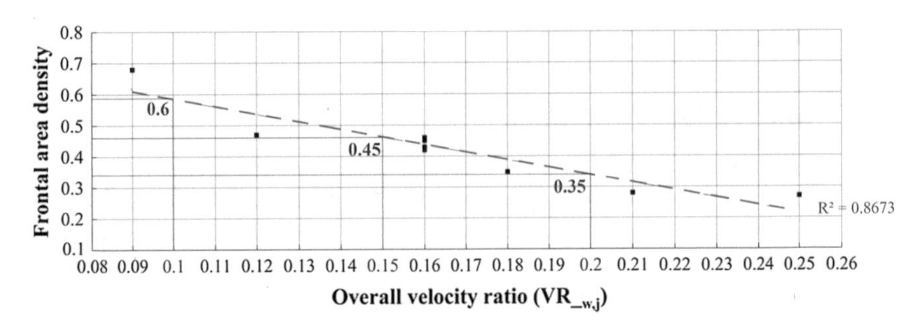

Fig. 3.1 Linear relationship between $\lambda_{f(z)}$ and $VR_{w,j}$. Values of $\lambda_{f(z)}$ were classified as: (1) $\lambda_{f(z)} \leq 0.35$, (2) $0.35 < \lambda_{f(z)} \leq 0.45$, (3) $0.45 < \lambda_{f(z)} \leq 0.6$, and (4) $\lambda_{f(z)} > 0.6$

stated in the following sections, i.e., classification of $\lambda_{f(z)}$; height increment z, and local prevailing wind probability P_θ.

3.4 Modeling Settings

3.4.1 *Classification of* $\lambda_{f(z)}$

Based on the linear relationship reported in Chap. 2, values of $\lambda_{f(z)}$ were classified as follows: (1) $\lambda_{f(z)} \leq 0.35$, (2) $0.35 < \lambda_{f(z)} \leq 0.45$, (3) $0.45 < \lambda_{f(z)} \leq 0.6$, and (4) $\lambda_{f(z)} > 0.6$, as shown in Fig. 3.1. This classification aims to statistically weigh the effects of different values of $\lambda_{f(z)}$ on the pedestrian-level natural ventilation performance, and to detect the potential air paths (the areas with low surface roughness) in high-density urban areas. For instance, Class 4 ($\lambda_{f(z)} > 0.6$) indicates that the wind velocity ratio ($VR_{w,j}$) may be less than 0.1, which implies very poor natural ventilation. In contrast, Class 1 ($\lambda_{f(z)} \leq 0.35$) indicates that $VR_{w,j}$ may be larger than 0.2, which implies good natural ventilation (Ng et al. 2011).

3.4.2 *Height Increment "z"*

To identify the "*z*" value, particularly in the context of Wuhan, we calculated the dividing level (27 m) of the building height distribution (0–204 m) using the local 3D building database in GIS. Building height data were analyzed in ArcGIS to identify the natural breakpoint (26.65 m), which classified the buildings at metropolitan area into the normal buildings (0–27 m) and high-rise buildings (27–204 m). As shown in Fig. 3.2, the percentage of normal building class is significantly larger than that of high-rise building class. Consequently, the urban morphology density at the layer

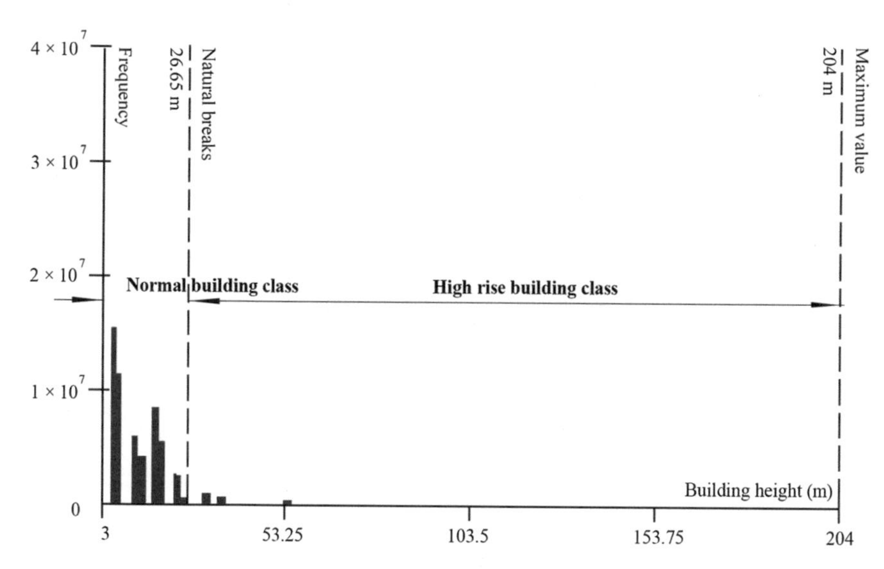

Fig. 3.2 Distribution of building height at metropolitan area of Wuhan. Natural breakpoint was identified at 26.65 m

ranging from 0 to 27 m is considered as being much higher than at the 27–204 m layer. Therefore, the "z" value for Wuhan city is set to 27 m.

3.4.3 *Local Prevailing Wind Probability P_θ*

To identify the local annual prevailing wind probability P_θ of Wuhan, the wind frequency data from Wuhan Observatory was used. The prevailing wind directions were identified as south ($\theta_1 = 90°$), southeast ($\theta_2 = 135°$), and southwest ($\theta_3 = 45°$), and their frequency was generally similar. Therefore, the values of $P_{\theta,i}$ ($i = 1, 2, 3$) across the three prevailing wind directions are simplified as 1/3.

3.5 Modeling Results

After calculating $\lambda_{f(z)}$ at a resolution of 100 × 100 m and classifying the results based in Fig. 3.1, the urban permeability of pedestrian-level natural ventilation in Wuhan was mapped, as shown in Fig. 3.3. Given the uncertainties in the modeling results caused by linear regression analysis and other assumptions, the modeling results are considered to be acceptable for the planning practices in the initial stages of decision-making process.

Fig. 3.3 Urban permeability map of the pedestrian-level natural ventilation in Wuhan (For interpretation of the references to color in the text, the reader is referred to the electronic version of this book)

3.6 Implementation in Urban Planning

The urban permeability map shown in Fig. 3.3 provides urban planners with an intuitive grasp of the natural ventilation of urban areas for the master planning in which the district land use and density are determined. The lower urban permeability areas (Classes 3 and 4), which include Hankou, Wuchang, and Hanyang, the downtown areas of Wuhan, indicate that the airflow in the street canyon is seriously restricted by compact building blocks, thereby increasing the risk of worsening the outdoor natural ventilation in these areas. In contrast, the surface roughness in other districts located far from the downtown area is very low $\lambda_{f(z)} \leq 0.35$). Compared with those districts where new development is still acceptable, the urban density at Hankou, Wuchang, and Hanyang should be strictly controlled in the master plan, and particular mitigation strategies in the district planning for these areas are also necessary.

Based on the above analysis, district-based information is needed to identify the district planning goals and mitigation strategies. High-resolution urban permeability maps are shown in Fig. 4.4a, b. The areas with low urban permeability occupy much of Hankou, compared to those areas in Hanyang and Wuchang. Furthermore, the areas with low urban permeability in Hankou are wide and adjacent to each other, whereas those areas in Wuchang and Hanyang are smaller and scattered apart. Thus, different district planning goals and mitigation strategies should be suggested in the respective districts.

3.6.1 *Planning Goals and Mitigation Strategies for Hankou*

The planning goal for Hankou district is to identify key areas to make potential air paths that can facilitate the movement of fresh air into the deeper urban areas of the city. This strategy is considered to be more practical than the strategy to decrease the urban density of the whole district that has wide urban area but low permeability. Based on the planning goal, the corresponding planning strategies are as follows.

First, as shown in Fig. 3.4a, the areas with low permeability are marked by white-dashed line boundaries. The gaps between these boundaries, represented by blue hollow arrows, are characterized by comparatively low surface roughness and are considered to be the key areas for potential air path I. The ground coverage ratio (λ_p) in these key areas needs to be strictly controlled to ensure that air paths are connected to each other—that is, λ_p must be less than 30% (Yoshie et al. 2008). The width of the air path I can range from several hundred meters to one kilometer.

Second, air path II is located at the neighborhood scale, which is detected inside individual low-permeability areas and is represented by blue-dashed arrows in Fig. 3.4b. The potential air path II is important to divide single and wide low-permeability areas into smaller ones, so that the air can flow into and thereby mitigate the high intensity of urban heat island in these areas. The width of air path II is about 100 m. The λ_p values at the air paths need to strictly be kept below 30% (Yoshie et al. 2008).

3.6.2 *Planning Goals and Mitigation Strategies for Wuchang and Hanyang*

The low-permeability areas in Wuchang and Hanyang are scattered and smaller in size than those in Hankou. As a result, the planning goal for these two districts is to decrease the urban density of the whole district in order to prevent the further spread of the small and scattered low-permeability areas. No mitigation strategies are feasible in creating air paths in these two districts. Based on this planning goal, the corresponding district planning strategy is stated as: the land use density of new development projects needs to be controlled by the ground coverage ratio (λ_p), which must be less than 50%, or better yet less than 30%.

3.7 Conclusions

In wind tunnel experiments and CFD simulations, the air in the street canyon is treated as the control volume. As a viable and effective alternative, the morphological method is an empirical model based on the relationship between urban morphology parameters and experimental wind data. Because of this characteristic, the complicated

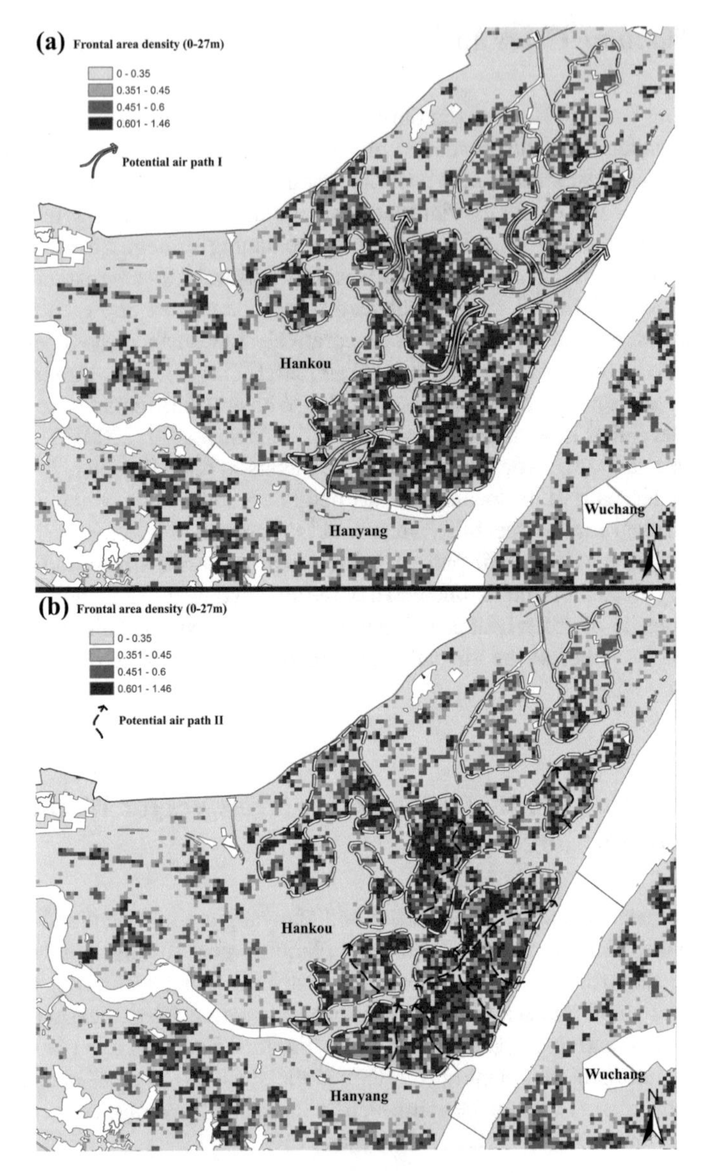

Fig. 3.4 **a** Potential air path I (urban scale). White-dashed lines marked the boundaries of areas with low urban permeability in Hankou. Potential air path I was represented by blue hollow arrows (For interpretation of the references to color in the text, the reader is referred to the electronic version of this book). **b** Potential air path II (neighborhood scale). White-dashed lines marked the boundaries of areas with low urban permeability in Hankou. Potential air path II was represented by blue-dashed line arrows (For interpretation of the references to color in the text, the reader is referred to the electronic version of this book)

calculations associated with fluid mechanics can be avoided during the planning process. Urban planners can readily apply the urban natural ventilation knowledge to the urban planning parameters, by using the local 3D building database.

The case study of Wuhan highlights practical application of the morphological modeling method and the establishment of planning guidelines for both master and district planning. As our knowledge of roughness parameters improves and more experimental data becomes available, the newly developed morphological modeling method will be more refined, and have a great potential for broader applications.

References

Architectural Institute of Japan (AIJ) (2008) National Research Project on Kaze-no-michi: making the best use of the cool sea breeze. Newsletter on Urban Heat Island Countermeasures, vol 4

Baumueller J, Hoffmann U, Reuter U (2009) Climate booklet for urban development—references for urban planning. Ministry of Economic Affairs Baden-Württemberg

Bottema M (1996) Roughness parameters over regular rough surfaces: experimental requirements and model validation. J Wind Eng Ind Aerodyn 64(2–3):249–265

Cionco RM (1965) Mathematical model for air flow in a vegetative canopy. J Appl Meteorol 4:517–522

Coceal O, Belcher SE (2004) A canopy model of mean winds through urban areas. Q J Roy Meteor Soc 130(599):1349–1372

Lettau H (1969) Note on aerodynamic roughness-parameter estimation on the basis of roughness-element description. J Appl Meteorol 8:828–832

MacDonald RW, Griffiths RF, Hall DJ (1998) An improved method for the estimation of surface roughness of obstacle arrays. Atmos Environ 32(11):1857–1864

Ng E (2012) Towards a planning and practical understanding for the need of meteorological and climatic information for the design of high density cities—a case based study of Hong Kong. Int J Climatol 32:582–598

Ng E, Yuan C, Chen L, Ren C, Fung JCH (2011) Improving the wind environment in high-density cities by understanding urban morphology and surface roughness: a study in Hong Kong. Landscape Urban Plann 101(1):59–74

Ren C, Ng E, Katzschner L (2011) Urban climatic map studies: a review. Int J Climatol 31(15):2213–2233

Yoshie R, Tanaka H, Shirasawa T, Kobayashi T (2008) Experimental study on air ventilation in a built-up area with closely-packed high-rise building. J Environ Eng 627:661–667(in Japanese)

Part II
Neighborhood Scale Wind Environment

Chapter 4
Semiempirical Model for Fine-Scale Assessment of Pedestrian-Level Wind in High-Density Cities

4.1 Introduction

This chapter introduces a fine-scale morphological modeling–mapping approach that provides pedestrian-level wind information of areas in between buildings and enables a more efficient decision-making in urban planning and design. This approach not only avoids high computational costs as opposed to what computational fluid dynamics (CFD) simulation requires (Tominaga et al. 2008; Frank 2006) but it also increases the resolution of the wind environment map to several meters, compared with hundred meters resolution in morphological models that are introduced in Part I. As a result, this new approach could bridge the gap between the current modeling methods and the requirements of urban design at neighborhood scale.

This chapter first summarizes the conventional modeling methods that are commonly used to estimate the urban wind environment, and then goes over the development of a new morphological index, which provides the high spatial resolution information about airflow between buildings without increasing the computational cost. The balance between momentum transfer and drag force is discussed in both an averaged sense over an area and a moving air parcel. In addition to the index, linear regression models are developed by correlating the new index with fine-scale wind tunnel experiment results and by validating the capability of the model to estimate real-life scenarios with different urban densities. The application of regression mapping approach to map the fine-scale wind environment, as well as two case studies to illustrate the different models of implementation in urban design, is provided toward the end of the chapter.

Originally published in Chao Yuan, Leslie Norford, Rex Britter and Edward Ng, 2016. A modelling-mapping approach for fine-scale assessment of pedestrian-level wind in high-density cities. Building and Environment, 97, pp. 152–165, © Elsevier, https://doi.org/10.1016/j.buildenv.2015.12.006.

C. Yuan, *Urban Wind Environment*, SpringerBriefs in Architectural Design and Technology, https://doi.org/10.1007/978-981-10-5451-8_4

4.2 Literature Review

The flow and dispersion in urban areas can be spatially classified into regional (up to 100 or 200 km), urban (up to 10 or 20 km), neighborhood (up to 1 or 2 km), and street scale (less than ~100 to 200 m), with different regimes of applicability and accuracy for each modeling method (Britter and Hanna 2003). Using a combination of numerical models in different scales, one can study outdoor urban aerodynamic properties. Regional- and urban-scale numerical models, e.g., the Mellor–Yamada hierarchy model (Mochida et al. 1997; Murakami 2004), the Fifth-Generation NCAR Mesoscale Model (MM5) (Grell et al. 1994), and the Weather Research and Forecasting model (WRF) (Skamarock et al. 2005), can provide information of the input boundary conditions for a CFD simulation at the neighborhood and street scales (Letzel et al. 2008; Blocken et al. 2012). CFD has been the most common design tool for planning and design practices (Murakami 2006). However, CFD requires high computational cost in the simulation and more processing time (e.g., months), especially when several input wind directions are considered, to calculate the averaged pedestrian-level wind speed (Yuan and Ng 2014). Also, given that any urban design options are often modified during the planning process, replicating the lengthy CFD modeling procedure would slow down the whole process. Thus, CFD is not suitable for the quick design and planning processes, and a more practical modeling is needed.

Since the current numerical modeling methods do not satisfy the requirements in practical urban planning and design (e.g., rapid but comprehensive evaluation), Ng et al. (2011) suggests that empirical morphological modeling may be a more practical tool for urban planning, as indicated in Part I. Empirical morphological methods are developed by correlating geometric indices [e.g., frontal area density (the ratio between frontal area and site area, λ_f) and the site coverage ratio (the ratio between built area and site area, λ_p)] with experimental wind data, so that the algorithms of geometric indices can be used to estimate the aerodynamic properties, such as roughness length (z_0) and displacement height (z_d) , which parameterize the effect of buildings on air flow. It was proved that these geometric indices are strongly related to pedestrian-level wind speed (U_p) normalized by the wind speed at a reference height ($U_{reference}$), i.e., the wind velocity ratio (VR) (Ng et al. 2011; Kubota et al. 2008). Consequently, the values of these geometric indices in pixels, commonly calculated in Geographic Information System (GIS) (Gál and Unger 2009; Ng et al. 2011; Wong et al. 2010; Yim et al. 2009; Yuan et al. 2014), can evaluate the wind environment in spatially continuous heterogeneity urban areas. Therefore, the conventional semiempirical morphological models enable quick decision-making in land use and urban density at the early stage of urban planning. Using the morphological models, the complicated calculations of fluid mechanics can be avoided, thereby significantly reducing the computational costs.

Bentham and Britter (2003) developed a practical and comprehensive model to estimate the uniform wind speed (U_c) at the urban canopy layer. The range of λ_f is from 0.3 to 0.01, and the experiment data of U_c normalized by frictional velocity

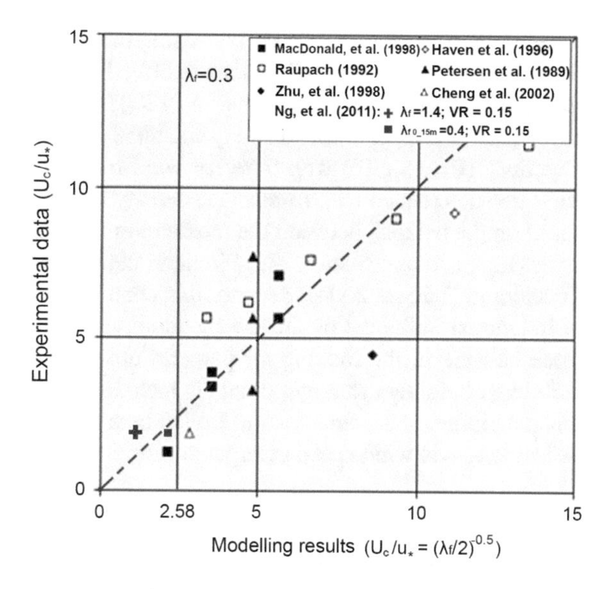

Fig. 4.1 Comparison of morphological modeling results for U_c with experiment data (Bentham and Britter 2003, edited by authors). The comparison result for the high-density area is included. The value of U_c/u_* of experimental data is estimated by the wind experimental data at Hong Kong

u_* is well related with modeling result, as shown in Fig. 4.1. Given the local values of λ_f and U_c/u_* which were estimated by the wind speed data from wind tunnel experiment (i.e., λ_f = 1.4 and VR = 0.15), we plotted Hong Kong data (Ng et al. 2011) in Fig. 4.1. The results indicate that Bentham and Britter's model is still valid with λ_f equal to 1.4. Figure 4.1 also collectively shows the low wind speed at the urban canopy layer and high-density obtained in this study, as compared with the modeling results from previous studies (MacDonald et al. 1998; Raupach 1992; Zhu et al. 1998; Havens et al. 1996; Petersen and Ratcliff 1989; Cheng and Castro 2002).

Despite the strength of the above-mentioned models, they are not suitable for estimating wind speed at the neighborhood-scale, i.e., the order of 1 km horizontal spacing (Belcher et al. 2003). These models can only provide spatially averaged wind information, with resolutions ranging from several hundred to thousand meters, since they are developed based on the balance between spatially averaged momentum fluxes and drag force. However, the mean wind speed between buildings significantly varies at the neighborhood scale due to spatial heterogeneity, and is important for evaluations of outdoor thermal comfort and air quality (Coceal and Belcher 2005). As a result, a more refined morphological model is needed for applications at the neighborhood scale.

4.3 Development of a Morphological Model

In order to address the limitations of both CFD simulation and large-scale morphological models on the application at the neighborhood scale, this chapter first introduces a refined morphological model to estimate the pedestrian-level wind speed in high resolution without increasing the computational cost. The model was based on developing approximations to directly avoid calculating the flow between roughness elements. As Belcher et al. (2003) did, airflow is assumed in a fully developed atmospheric boundary layer annually blow through urban areas in 16 wind directions (i = 1,…,16) with different wind frequencies (P_i). The pedestrian-level wind speed is related with frontal area density (λ_f) over a certain height range (Δz), based on the balance between the momentum transfer and drag force in the canopy layer (Raupach and Shaw 1982; Finnigan 2000, 1985; Bentham and Britter 2003; Coceal and Belcher 2004). Second, this chapter particularly focuses on the effect of individual buildings on airflow by addressing the momentum transfer and drag force balance in the moving air parcel. Subsequently, a distance index (L) from individual buildings to target points is included in the morphological model. With this new index, the point-specific frontal area density is developed to evaluate the pedestrian-level wind speed at target points.

4.3.1 Relating λ_f to the Pedestrian-Level Wind Speed at High-Density Areas

Given a steady and uniform airflow, there is a balance between the drag force of buildings on the air flow and turbulent momentum transfer downward from above as

$$\rho D = -\rho \frac{\partial}{\partial z} \langle u'w' \rangle , \tag{4.1}$$

where ρD, the left side of equation, is the canopy drag force, or specifically the body force per unit volume on the spatially averaged flow, and $\rho \langle u'w' \rangle$, the right side of equation, is the momentum flux caused by turbulent mixing. The momentum flux by turbulent mixing can be considered as shear stress (τ_w) that is given by Schlichting and Gersten (2000):

$$\frac{\partial}{\partial z} \tau_w = -\rho \frac{\partial}{\partial z} \langle u'w' \rangle \tag{4.2}$$

The total canopy drag force (ρD) (aka. body force per unit volume) is given by Bentham and Britter (2003):

$$\rho D = \frac{1}{2} \rho U_c^2 \frac{\sum_{\text{obstacle}} (C_D A_{\text{front}})}{h A_{\text{site}} (1 - \lambda_p)} , \tag{4.3}$$

where A_{front} is the frontal area, U_c is the averaged wind speed in the canopy layer, and C_D is the drag coefficient. The air volume is given by $h A_{\text{site}}(1 - \lambda_p)$, where A_{site} is the site area, h is the canopy height, and λ_p is the site coverage ratio. Substituting the expression for $\rho \frac{\partial}{\partial z} \langle u'w' \rangle$ (Eq. 4.2) and ρD (Eq. 4.3) into Eq. (4.1) yields another statement of the balance between canopy drag and vertical transfer of horizontal momentum in height z is as follows:

$$\frac{\partial}{\partial z}\tau_w = \frac{1}{2}\rho U_{(z)}^2 \sum\nolimits_{\text{obstacle}} \left(C_{D(z)} \frac{dA_{\text{front}}}{dz} \right) / A_{\text{site}}(1-\lambda_p) \tag{4.4}$$

It should be noted that the right side of Eq. 4.4 is the sectional drag acting only at height z, where U_z is the wind speed at height z and $C_{D(z)}$ is the sectional drag coefficient (Coceal and Belcher 2004). Cheng and Castro (2002) found that $C_{D(z)}$ is equal to 2.0 near the top, because air can flow over and around roughness elements at the top; since the air only can flow around roughness elements in the remaining depth, i.e., the sectional drag is enhanced, $C_{D(z)}$ increases to 3.0 over the remaining depth. Given the deep street canyon, $C_{D(z)}$, is equal to 3.0, the wind speed (U_i) in the ith layer of the street canyon is given by rearranging Eq. (4.4) as

$$U_i = \left(\frac{\tau_{w,i}}{3\rho} \cdot \frac{2(1-\lambda_p)}{\lambda_{f,i}} \right)^{0.5}, \left(\text{where } \lambda_{f,i} = \frac{A_{\text{front},i}}{A_{\text{site}}} \right) \tag{4.5}$$

Bentham and Britter (2003) ignored the vertical variation of $\tau_{w,i}$ using the equivalent surface shear stress (τ_w) and considered C_D is equal to 1.0. They treated the air volume between two building rows as the control volume; therefore, λ_p in Eq. 4.5 is equal to 0. Consequently, they provided a practical method to estimate the averaged velocity in the urban canopy U_c as

$$\frac{U_c}{u_*} = \left(\frac{\lambda_f}{2} \right)^{-0.5}, \left(\text{where } u_* \text{ is equal to } \left(\frac{\tau_w}{\rho} \right)^{0.5} \right) \tag{4.6}$$

Equation (4.5) relates the wind speed (U_i) to the corresponding values of $\lambda_{f,i}$, ignoring the vertical variation of $\tau_{w,i}$ as Bentham and Britter (2003) did. This understanding is consistent with MacDonald et al. (1998): the frontal area density above the displacement height (z_d) , λ_f^*, can better estimate the wind profile than λ_f. On the other hand, the near-ground wind speed is postulated to depend on λ_f', the frontal area density below z_d, instead of λ_f^*. In Chaps. 2 and 3, the layer is defined as 15 and 23 m in Hong Kong and Wuhan, respectively, due to the podium morphological characteristic and building height distribution. It is considered that the air is impeded at this layer as much as if it is under z_d. The wind velocity ratio (VR) is well correlated to λ_f', which is represented by $\lambda_{f(0-15\,m)}$, the frontal area density in the layer, rather than conventional λ_f, as shown in Fig. 4.2. Besides, the averaged results for the conventional λ_f in Hong Kong (Ng et al. 2011), $\lambda_f = 1.4$ and VR = 0.15, are consistent with the Bentham and Britter (2003)'s modeling result, as shown in Fig. 4.1.

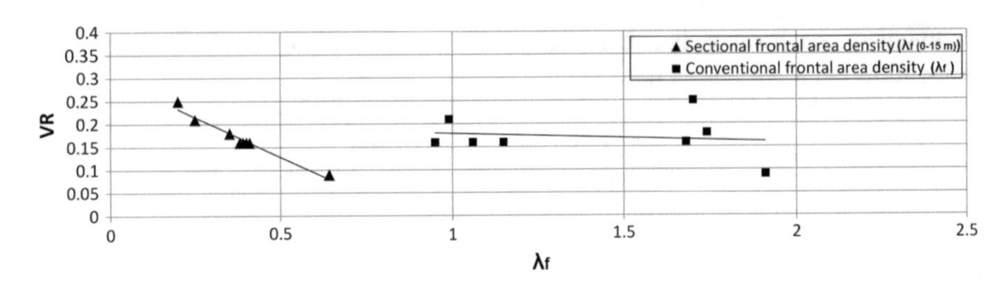

Fig. 4.2 Comparison of conventional and sectional λ_f. VR and λ_f are the spatially averaged results in 300 × 300 m grid (For more details, please see Chap. 2)

4.3.2 *Adjustment of* λ'_f *to the Fine-Scale Wind Estimation*

It is assumed that the fine-scale wind speed between buildings can be modeled by adjusting the morphological index λ'_f using the distance from buildings to target points. As mentioned in the above section, the momentum transfer and drag force balance are spatially averaged across the whole control volume, and each building (i.e., roughness element) is treated equally to calculate the spatially averaged indices. This section illustrates how air flows through urban areas by describing the momentum transfer and drag force balance in a moving air particle. The air particle is impeded by the drag force of buildings and is accelerated by the downward transfer of horizontal momentum. This means that the air particle will stop without vertical transfer of horizontal momentum (Belcher et al. 2003; Hall 2010). The speed of the air particle can be recovered after traveling further away from the building. This indicates when the wind speed at test points is affected by the surrounding area, the effects of frontal area units are different given the different distances from the target points to individual frontal area units. Consequently, a new point-specific frontal area density $\left(\lambda_{\text{f\_point}}\right)$ can be developed by adjusting λ'_f using the distance index (L). This adjustment of calculation indicates that $\lambda_{\text{f-point}}$ is not only spatially and annually averaged but also point-specific. As shown in Fig. 4.3, frontal area pixels in Δz (0–15 m) $\left(A_{\Delta z,i}\right)$, weighted by the distance coefficient and annual wind frequency (P_i) in the ith wind direction, are added up and normalized by the scanned area to calculate the point-specific frontal area density $\left(\lambda_{\text{f\_point}}\right)$ as

$$\lambda_{\text{f\_point}} = \frac{\iint_D \mathrm{l}^c_{x,y}(A_{\Delta \mathrm{z,x,y}}/A_\mathrm{t})\mathrm{dxdy}}{A_\mathrm{t}}, \quad D = \{x^2 + y^2 \le r^2\}c \tag{4.7}$$

$$A_{\Delta z,x,y} = \sum_1^{i=16} A_{\Delta z,i}\,p_i \tag{4.8}$$

$$l_{\mathrm{x,y}} = \frac{r - L}{r} \tag{4.9}$$

$$A_\mathrm{t} = \pi \cdot r^2, \tag{4.10}$$

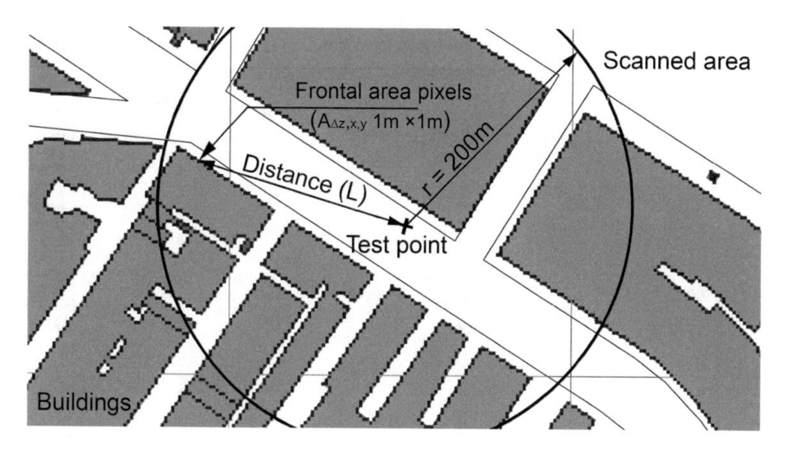

Fig. 4.3 Schematic illustration of λ_{f_point} calculation method. Test point, scanned area (A_t), radius (r), distance (L), and frontal area units at the boundaries of buildings are presented

where $A_{\Delta z,x,y}$ is the wind frequency-weighted frontal area at the pixel (x, y), in which x and y are the coordinates (the test point is the origin of the coordinate), Δz is from 0–15 m, and $l_{x,y}$ is the distance coefficient. A_t is the scanned area and r is the scan radius (200 m), as shown in Fig. 4.3. Belcher et al. (2003) indicated that the increase of the mean wind speed is not linear when the air particle gets further away from the roughness elements. Therefore, an exponent c, 2.0, is used to adjust $l_{x,y}$. As indicated in Table 4.1, the modeling results with the exponent $c = 2.0$ fit the experimental data better than the results obtained using the exponent $c = 1.0$.

4.4 Development of Regression Model

Point-specific index (λ_{f_point}) is correlated with wind tunnel data to test the model performance. Since VR were measured in a wind tunnel experiment conducted by the Hong Kong Planning Department (HKPD) (2005), we calculated the corresponding values of λ_{f_point} at test points using ArcGIS, in which the pixel values of $A_{\Delta z,x,y}$ and $l_{x,y}$ were calculated using the building geometry data from HKPD, and annual wind frequency data (P_i) was obtained from MM5/CALMET (California Meteorological) system (Yim et al. 2007). λ_{f_point} values were then entered into a linear regression analysis to predict the corresponding annually averaged wind VR.

The wind tunnel experiment was carried out at six locations across the territory, as shown in Fig. 4.4. Three of them (i.e., Mong Kok, Sheung Wan, and Causeway Bay) are located at the metropolitan area, and the remainder (i.e., Tuen Mun, Sha Tin, and Tseung Kwan O) are located in new town areas. Data from metropolitan areas were used to develop the linear regression model, whereas data from new town

Table 4.1 Slope coefficient, intercept, and standard error in regression equations for different districts. The result of sensitivity test of exponent c was also presented

Districts	Zones	R^2 (exponent $c = 1.0$)	R^2 (exponent $c = 2.0$)	Slope coefficient	Intercept	SE of predicted VR	SE of predicted U_p (m/s)
Mong Kok (Grid plan)	Zone a	0.52	0.56	−3.9	0.4	0.03	0.20
	Zone b	0.14	0.42	−3.8	0.4	0.04	0.27
Sheung Wan (Irregular street grid)	Zone a	0.35	0.49	−1.7	0.3	0.05	0.33
	Zone b	0.57	0.62	−1.7	0.3	0.04	0.27
Causeway Bay (Irregular street grid)	Zone a	0.47	0.49	−1.9	0.3	0.03	0.20
	Zone b	0.40	0.40	−1.6	0.3	0.04	0.27

Note Annually averaged wind speed at the reference height (500 m above the ground) at study areas is equal to 6.67 m/s, which is from MM5 modeling result (Yim et al. 2007)

areas were used to test the performance of the linear regression model at the urban areas with different densities.

4.4.1 Regression Analysis for High-Density Urban Areas

Each location at metropolitan areas had two zones (i.e., Zone a and Zone b), and test points were evenly distributed to collect the overall wind VR, as shown in Fig. 4.5. The overall wind VR was used in this regression analysis. The overall wind VR quantitatively describes the annually averaged wind permeability in Technical Circular No. 1/06, the Air Ventilation Assessment (AVA) in Hong Kong (Hong Kong Development Bureau (HKDB) 2006), and is defined as

$$\mathrm{VR} = \sum\nolimits_{i=1}^{16} P_i \cdot \mathrm{VR}_{500,i}, \tag{4.11}$$

where $\mathrm{VR}_{500,i}$ represents the directional wind velocity ratio between wind speeds at the pedestrian level and 500 m above the ground (reference height). Results of the regression analyses were plotted in Fig. 4.6a, b and summarized in Table 4.1. In general, the point-specific $\lambda_{\mathrm{f\_point}}$ was negatively associated with wind VR. In other

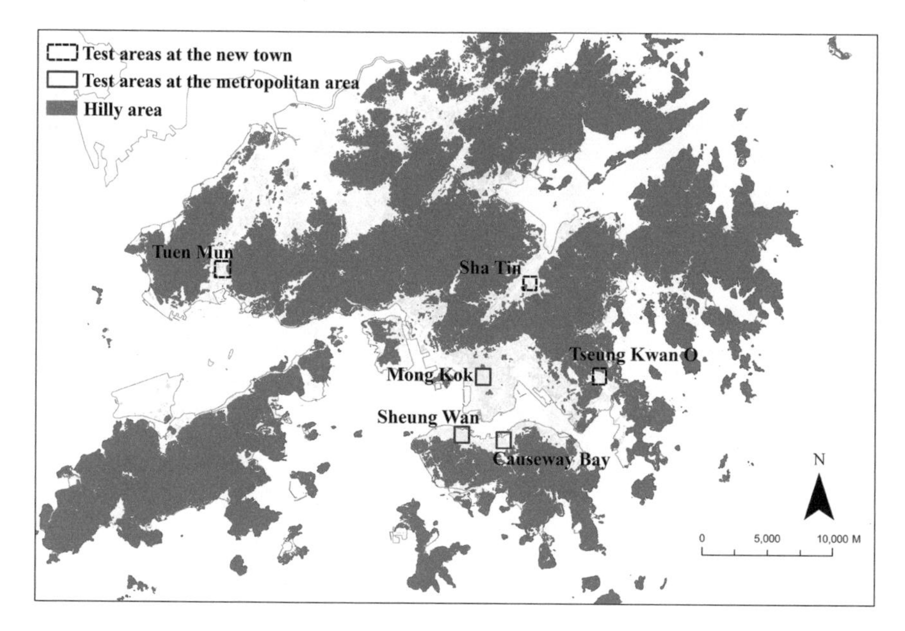

Fig. 4.4 Six test locations where the wind tunnel data was collected: black dash boxes: new town areas; red solid boxes: metropolitan areas

words, the closer the surrounding roughness elements to the test point, the more theairflow would be impeded, and the slower the pedestrian-level wind speed at a particular point would result.

Figure 4.6a, b shows that the slopes between street grids (regular versus irregular) differ. The slope for Mong Kok (−3.9) is significantly different from those for Sheung Wan and Causeway Bay (−1.7). The incoming air in areas with regular street grids might encounter strong resistance, and the wind speed is particularly small in the streets that are perpendicular to the wind direction, given that airflow at those streets is only driven by the limited horizontal momentum transferred from the street that is aligned with the wind direction. In contrast, the incoming air can easily flow around a building block in areas with irregular street grids, driven by the horizontal momentum. Thus, the same frontal area can impede more airflow into the regular street grid than the irregular street grid, which explains the different coefficients and intercepts in urban areas with different street grids in Table 4.1. Table 4.1 also shows that the slopes and intercepts for different test zones within the same street grid are similar. Therefore, it is possible to develop the general regression equation to predict VR using λ_{f_point} for each type of street grids as shown below:

(i) for districts with regular street grids (main streets perpendicular with each other, as shown in Fig. 4.5):

$$VR = -3.9\lambda_{f\_point} + 0.41 \tag{4.12}$$

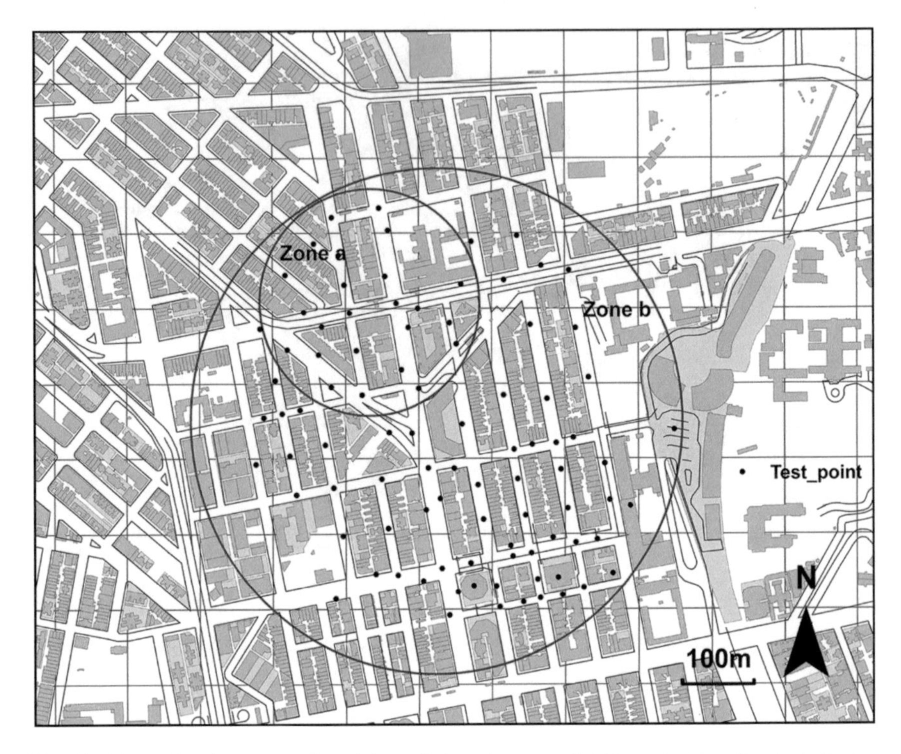

Fig. 4.5 Test points in test locations (Mong Kok as the example). Every test location includes two zones (Zone a and Zone b)

(ii) for districts with irregular street grids, as shown in Fig. 4.10

$$VR = -1.7\lambda_{f\_point} + 0.28 \quad (4.13)$$

4.4.2 Regression Analysis for Low-Density Urban Areas

In the above section, the regression equations were developed based on the data from the metropolitan areas of Hong Kong. But the relations between λ_{f_point} and VR could be different at urban areas with different densities, such as λ_f and λ_p. Therefore, the performance of the regression model at new town areas with lower densities was tested, as shown in Fig. 4.7. Different densities, representing by λ_p, λ_f and $\lambda_{f0-15\,m}$, at six test sites are shown in Table 4.2. It is evident that the urban density of new town areas (Fig. 4.7a, b, and c) is much lower than at metropolitan areas (Fig. 4.7d, e, and f).

The VR and λ_{f_point} from these three new town areas, together with data from the metropolitan areas, are plotted in Fig. 4.8. The linear relations for the low-density urban areas are consistent with those for the high-density areas, despite of

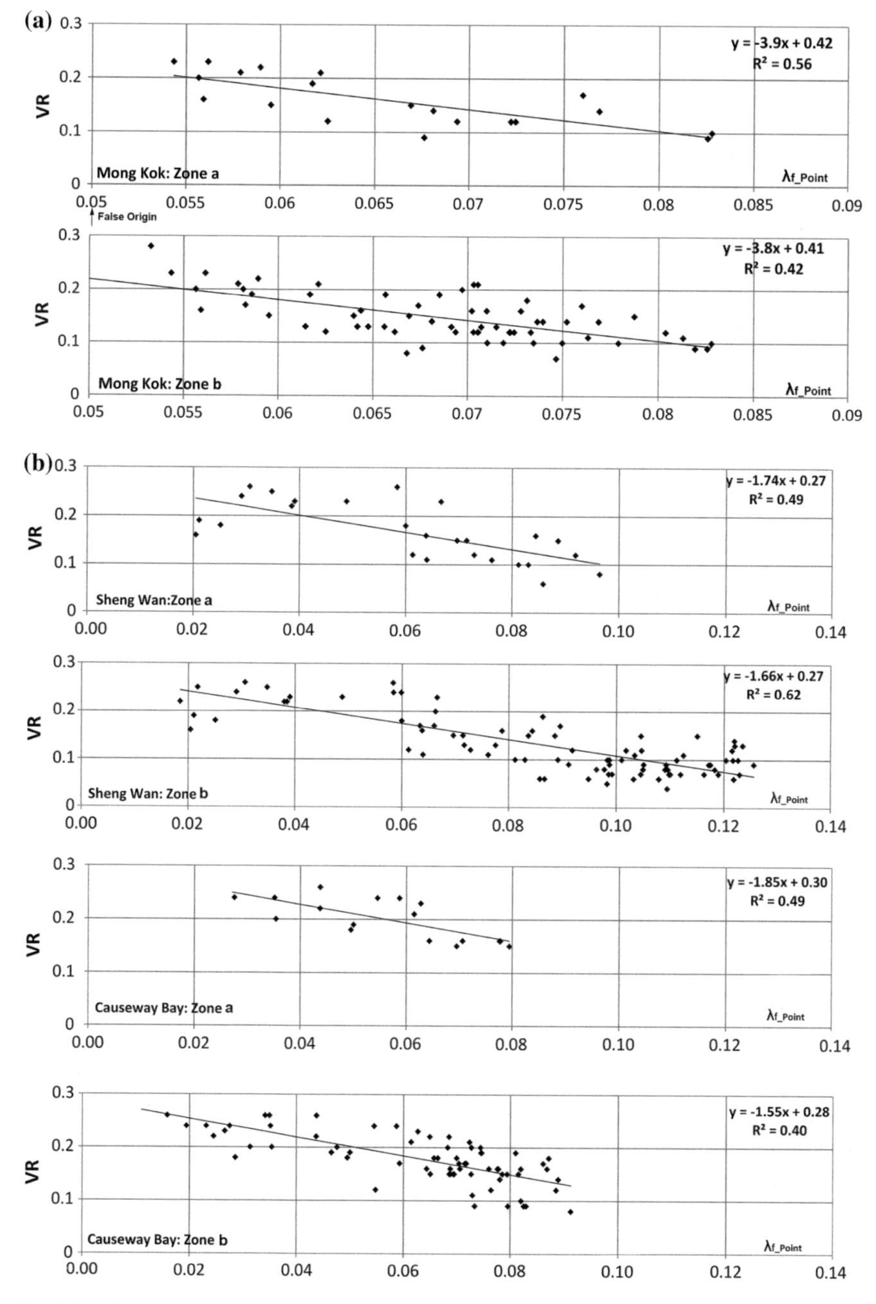

Fig. 4.6 **a** Linear regression analysis at Mong Kok: Zone a and Zone b. **b** Linear regression analysis at Sheung Wan and Causeway Bay: Zone a and Zone b

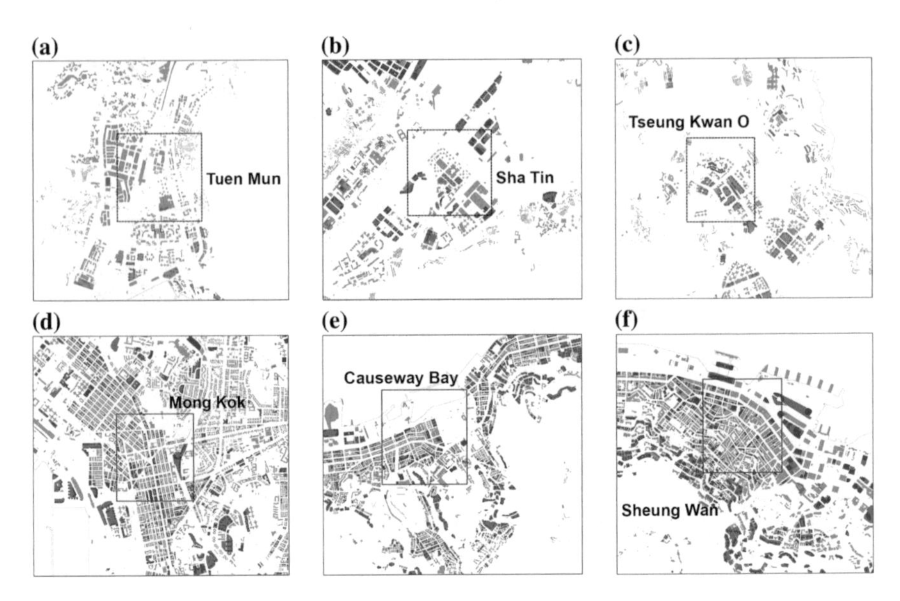

Fig. 4.7 Different densities between new town areas (**a**, **b**, **c**) and metropolitan areas (**d**, **e**, **f**). The calculating areas are highlighted

Table 4.2 Different densities at six test sites, represented by λ_p, λ_f and $\lambda_{f(0-15\,m)}$

	(a) Tuen Mun	(b) Sha Tin	(c) Tseung Kwan O
λ_p	0.27	0.22	0.18
$\lambda_{f(0-15\,m)}$	0.16	0.22	0.13
λ_f	0.39	0.86	0.55
	(d) Mong Kok	(e) Causeway Bay	(f) Sheung Wan
λ_p	0.4	0.32	0.45
$\lambda_{f(0-15\,m)}$	0.36	0.26	0.45
λ_f	1.00	1.09	1.68

the narrower range of λ_{f_point} at the low-density urban areas. Similarly, the linear regression equation is consistent with Eq. (4.13), which is applicable for an irregular street grid. This result indicates that the sensitivity of VR to the change of λ_{f_point} is independent from the background urban density; in another words, the same linear equation could be used at the different urban areas or cities. However, it should be noted that local urban planners should identify the corresponding value of Δz in the calculation of λ_f' (sectional λ_f), which is based on the local urban morphology as did in Chap. 2, before applying this model to their own cities. A $\lambda_{f(0-15\,m)}$ (Δz : $0-15\,m$) was applied in this study due to the specific podium morphological characteristic at Hong Kong, as shown in Fig. 2.5.

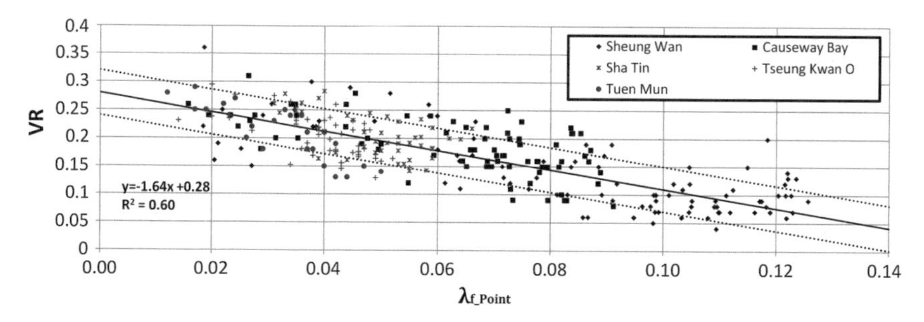

Fig. 4.8 Sensitivity analysis of new morphological model to determine pedestrian-level wind speed normalized by wind speed at the reference height. The black plots represent the urban areas with high density, whereas the red plots represent the urban areas with low density. The standard error is 0.04

4.4.3 Model Performance

The statistical analysis indicates an acceptable modeling–mapping method from the practical planning and design perspective. The values of R^2 ranged from 0.40 to 0.62. As shown in Fig. 4.9a, b, the 95% confidence intervals (CI) of the regression model slopes are -1.66 ± 0.14 and -3.83 ± 0.56 for irregular and regular street grids, respectively. The standard error (SE) of predicted VR at test zones is about 0.04. The CI and SE values are one order of magnitude smaller than the corresponding predicted values. Furthermore, with the 6.67 m/s annually averaged wind speed at the reference height (500 m above the ground) at the test locations obtained by the MM5 modeling result (Yim et al. 2007), an SE of 0.27 m/s for predicted pedestrian-level wind speed was calculated. The sensitivity of physiological equivalent temperature (PET) on the change of wind speed at this range is low. Therefore, it is considered that the accuracy of the modeling result is acceptable to the purpose of urban planning and design, i.e., the urban designer can use λ_{f_point} to appropriately predict VR.

4.5 Mapping Pedestrian-Level Wind Environment

Based on the above findings, λ_{f_point} was applied to map the pedestrian-level wind speed in the neighborhood scale. First, a self-developed program was used, which is embedded as a Visual Basic for Applications (VBA) script in the ArcGIS System, to calculate λ_{f_point} using Eqs. (4.7)–(4.10) on a pixel by pixel (1 m $\times$ 1 m) basis of all non-built-up locations in the target area. Given that the scan radius r was 200 m, a 200 m buffer bandwidth was required in the calculation area to ensure that the calculation of pixels within the target area boundary included the corresponding urban context. Second, the regression mapping approach was applied to extend λ_{f_point} map to the overall wind VR map using the regression either Eqs. 4.12 or 4.13.

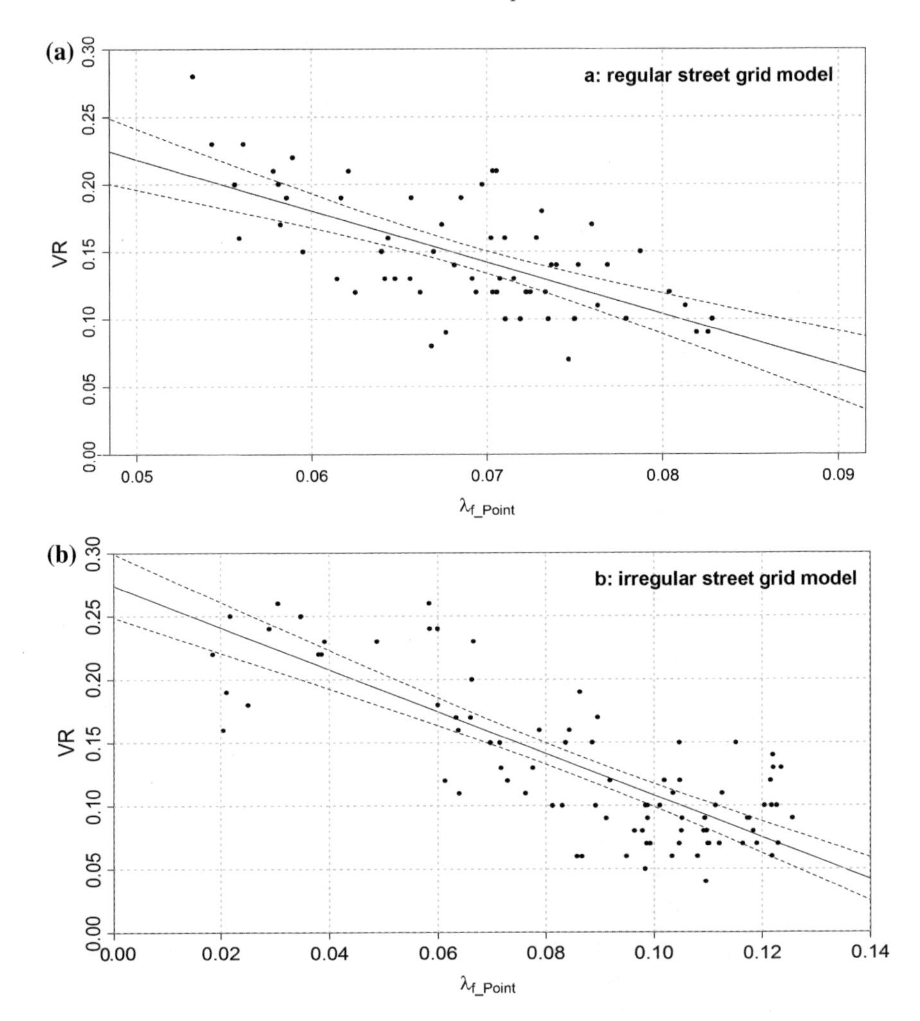

Fig. 4.9 **a** 95% confidence interval for regression analysis for the regular street grid. The slope is equal to -1.66 ± 0.14, taking Mong Kok (Zone b) as example. **b** 95% confidence interval for regression analysis for the irregular street grid. The slope is equal to -3.83 ± 0.56, taking Sheung Wan (Zone b) as example

The regression mapping approach has been broadly used to evaluate air pollutant conditions and predict the indoor temperatures of urban areas (Smargiassi et al. 2008). The regression mapping approach has the advantage of accuracy over the spatial interpolation modeling (Briggs et al. 1997) that requires numerous monitors to adapt to the complexity and diversity of urban areas.

Taking Sheung Wan and Central as an example, the results are shown in Fig. 4.10a (the map of λ_{f_point}, b (the map of VR). They quantitatively present the urban permeability at the target area. These maps reflect a common perception of the wind environment. Specifically, urban permeability is the highest (green area) in coastal areas (VR larger than 0.2). The wind permeability significantly decreases toward deep urban areas (VR smaller than 0.1) in prevailing wind directions (red area), given that the model took the wind frequency into account in the calculation of λ_{f_point}.

Based on the definition of wind VR (Eq. 4.11), annually averaged pedestrian-level wind speed (U_p) was also calculated and mapped by multiplying VR by U_{500} (annually averaged wind speed) at the reference height, 6.67 m/s at Sheung Wan and Central. Using the wind classification based on PET (Cheng et al. 2011), the neighborhood wind speed field can be classified into Class 1 (Poor, <0.6 m/s), Class 2 (Low, 0.6–1.0 m/s), Class 3 (Satisfactory, 1.0–1.3 m/s), and Class 4 (Good, >1.3 m/s). The pedestrian-level wind speed map with the classification is shown in Fig. 4.11. Both wind permeability map (Fig. 4.10) and pedestrian-level wind speed map (Fig. 4.11) provide useful urban design information without needing CFD and wind tunnel experiment.

4.6 Implementation

The new modeling and mapping method, with the low computational cost, accuracy, and high-resolution modeling results, is considered as a practical tool for the urban planning in the neighborhood scale. Compared to the several months, it may take for CFD simulation or wind tunnel experiment to obtain the similar annually averaged results as shown in Figs. 4.10 and 4.11; it only takes a few hours or at most days of processing time using our new modeling method. Also, our new method enables urban planners and designers to model and evaluate the wind environment by themselves at the early stage when the proposed design or planning still needs to be modified. However, it should be noted that the planners and designers should take the entire modeling results into account, rather than merely evaluate the values by pixel. For important projects with higher requirement for accuracy, CFD simulation or wind tunnel experiment is still required.

In the following section, the application of the new modeling–mapping approach in the air ventilation assessment (AVA) of Hong Kong is illustrated. According to the Technical Circular No. 1/06 (HKDB 2006), AVA composes of three steps: expert evaluation, initial study, and detailed study. Expert evaluation provides a qualitative assessment of design options and leads to the pilot or detailed study in which a quantitative evaluation is conducted. This new modeling–mapping approach can be applied in both the expert evaluation and pilot study to provide a general pattern of pedestrian-level wind environment without needing CFD simulation or wind tunnel experiment. Modeling results can be reported as the wind velocity ratio VR or even directly the pedestrian-level wind speed, if the local wind speed at the reference height data is available.

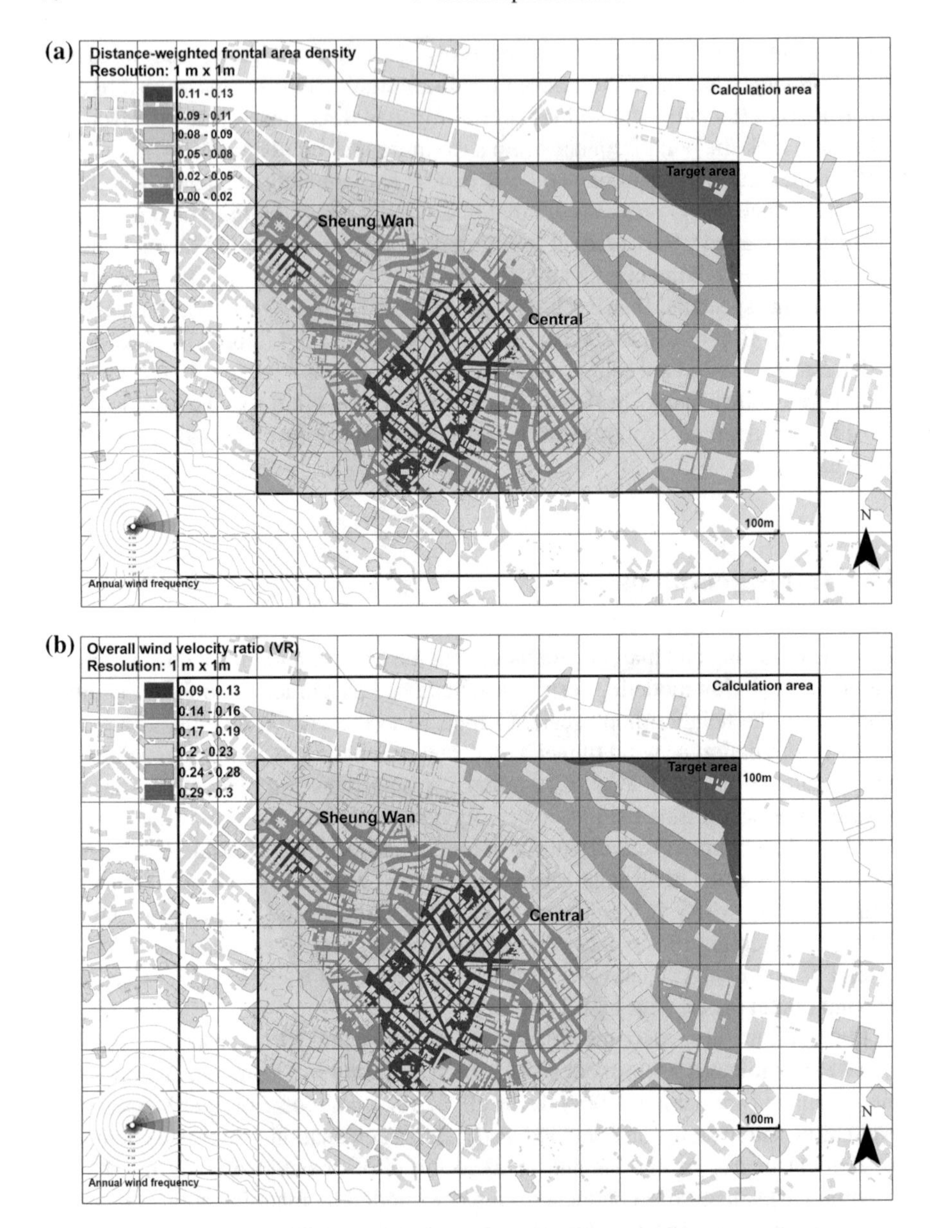

Fig. 4.10 **a** Map of $\lambda_{f\text{-point}}$ for Sheung Wan and Central. **b** Estimated overall wind velocity ratio (VR) map which represents the annually averaged wind permeability. Wind permeability in coastal areas is high (green area) and significantly decreases toward deep urban areas in prevailing wind directions (red area) (For interpretation of the references to color in the text, the reader is referred to the electronic version of this book)

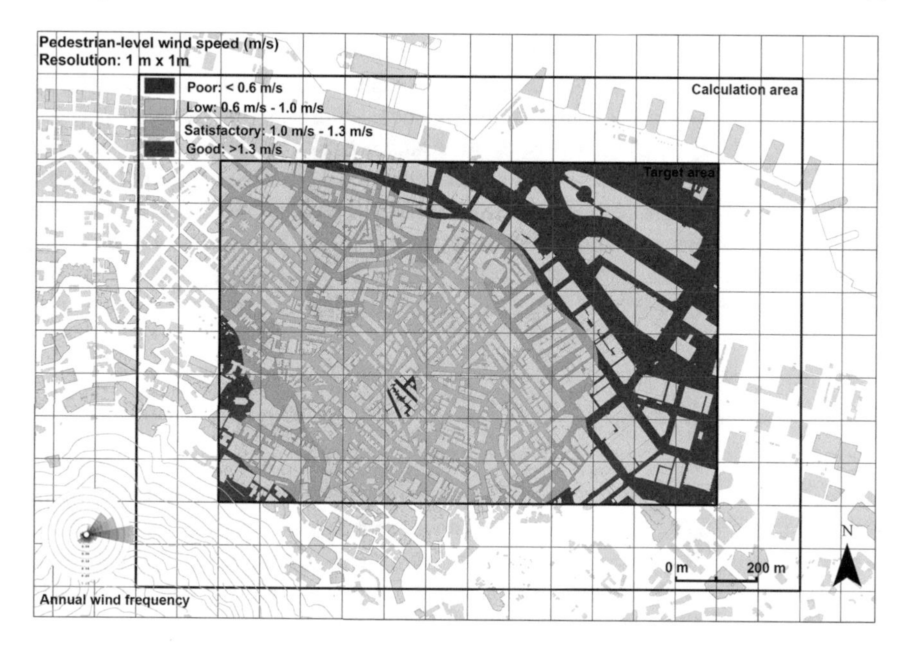

Fig. 4.11 Pedestrian-level wind speed map, which was classified as Class 1 (Poor, <0.6 m/s), Class 2 (Low, 0.6–1.0 m/s), Class 3 (Satisfactory, 1.0–1.3 m/s), and Class 4 (Good, >1.3 m/s) (For interpretation of the references to color in the text, the reader is referred to the electronic version of this book)

4.6.1 District Planning

Taking Sheung Wan and Central as an example, Chap. 2 has shown that the wind permeability in these two areas is low. Although useful in master planning, the spatial resolution of the wind permeability information is not sufficient enough to improve the local wind environment at the district planning stage. The new approach produces a pedestrian-level wind speed map (Fig. 4.11) that provides fine-scale wind information (e.g., wind speed between buildings) and bridges the gap between master planning and district planning. As shown in Fig. 4.11, compact building groups could impede airflow in the purple zone (Class 1: Poor, < 0.6 m/s) that was associated with greater outdoor thermal discomfort and poor air pollution dispersion. While the wind environment at the yellow zone (Class 2: Low, 0.6–1.0 m/s) is better than the one at the purple zone, the wind speed is still low and may not mitigate the outdoor thermal discomfort and air pollution concerns. In contrast, the wind speed in the light green zone is satisfactory (Class 3: Satisfactory, 1.0–1.3 m/s), which could improve natural ventilation performance and air quality. The dark green zone (Class 4: Good, >1.3 m/s), mostly located at the waterfront area with the lowest urban density, has the highest urban permeability.

Based on the above evaluation, urban planning and design strategies can be tailored to address specific wind environment issues at different zones. New development projects should be prohibited at the purple zone, and projects should be strictly controlled at the yellow zone with careful AVA evaluation. Given the low air pollutant dispersion in the purple and yellow zones, planners should not propose additional bus stops, terminus, heavy traffic roads, or other land use with pollutant sources at these zones to avoid trapping the emitted air pollutant. On the other hand, new projects can be developed in the light green zone with cautions to prevent impairment of existing good wind environment. Public waterfront parks, as well as new bus stops, terminus, and other traffic facilities, can be arranged at the dark green zone, given the outdoor thermal comfort and the rapid dispersion of traffic-related air pollution.

4.6.2 *Building Design*

Urban designers can apply this modeling–mapping method to evaluate the effect of the new building on the neighboring wind environment. Two scenarios were included in this case study: (a) the scenario without and (b) with new building. Figure 4.12a, b shows the new building, the surrounding street grid, and prevailing wind directions. In scenario a (Fig. 4.12a), the modeling results indicate poor existing wind environment surrounding the site, as depicted by the red area with VR less than 0.2, and better wind environment along Hennessy Road (i.e., the brown area with VR larger than 0.2). In scenario b when we put a new building into the site, the red area with lower VR is enlarged in both windward and leeward directions. The affected areas are highlighted in Fig. 4.12b.

4.7 Conclusion

Upon broadly discussing the aerodynamic properties of urban areas and the corresponding planning principles, we developed a modeling–mapping approach to estimate the pedestrian-level wind speed in the neighborhood scale. High spatial resolution modeling results and low computational cost are the main attractive features of this new approach. The key summaries from this chapter are as follows:

- Given the balance between vertical flux of horizontal momentum and horizontal drag force in the street canyon layer by layer, the sectional frontal area density $\left(\lambda_f'\right)$ below z_d was calculated to estimate the pedestrian-level wind speed, particularly in the high-density urban areas where the high buildings interfere with each other.
- λ_{f_point} was developed to estimate the effect of the neighborhood wind permeability on a particular point. λ_f' was modified by discussing the momentum transfer balance in a moving air particle. The distance index (L) was included into the morphological model to investigate the different effects of individual building on the wind speed at the target point.

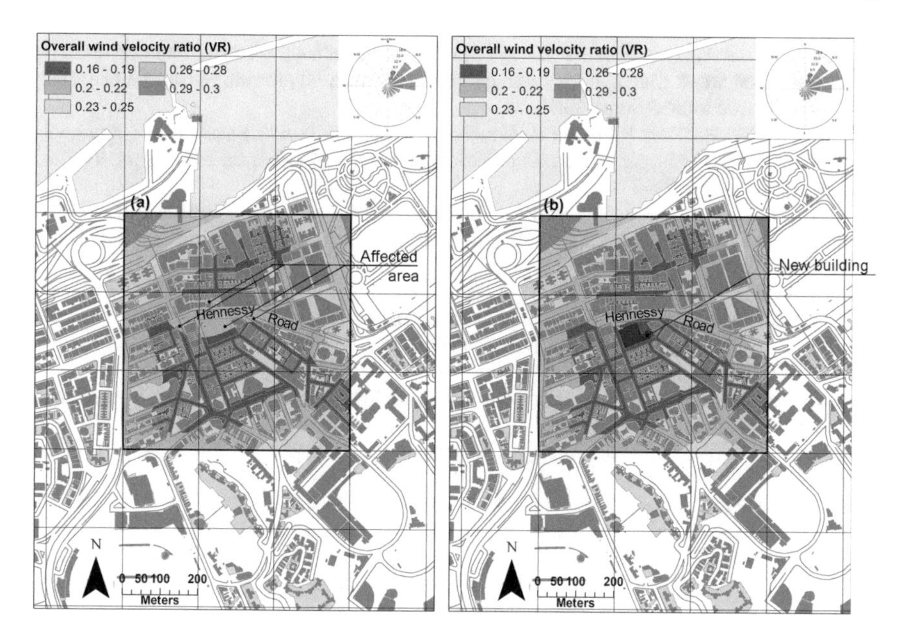

Fig. 4.12 Evaluation of the effect of the new building on the neighborhood wind environment. **a** Original scenario; **b** Scenario with the new building

- Accuracy of the new morphological model was tested using statistical analysis. Two regression equations were developed to form a semiempirical assessment tool to evaluate the pedestrian-level wind environment. The accuracy of the modeling results is acceptable for the urban design purpose at both high- and low-density urban areas.
- Various effects of regular and irregular street grids on the pedestrian-level wind speed, which are caused by different horizontal momentum transfer processes, were clarified. In the regular street, airflow at the streets perpendicular to the wind direction is only driven by the limited horizontal momentum transferred from the street that is aligned with the wind direction.
- New method was applied to two case studies in Hong Kong to illustrate how the modeling results could be used in the planning and design practice for better evidence-based decision-making at the early stage of city planning.

References

Belcher SE, Jerram N, Hunt JCR (2003) Adjustment of a turbulent boundary layer to a canopy of roughness elements. J Fluid Mech 488:369–398

Bentham T, Britter R (2003) Spatially averaged flow within obstacle arrays. Atmos Environ 37(15):2037–2043

Blocken B, Janssen WD, van Hooff T (2012) CFD simulation for pedestrian wind comfort and wind safety in urban areas: general decision framework and case study for the Eindhoven University campus. Environ Modell Softw 30:15–34

Briggs DJ, Collins S, Elliott P, Fischer P, Kingham S, Leberet E, Pryl K, Reeuwijk HV, Smallbone K, Veen AV (1997) Mapping urban air pollution using GIS: a regression-based approach. Int J Geogr Inf Sci 11(7):699–718

Britter RE, Hanna SR (2003) Flow and dispersion in urban areas. Annu Rev Fluid Mech 35:469–496

Cheng H, Castro IP (2002) Near wall flow over urban-like roughness. Bound-Layer Meteorol 104(2):229–259

Cheng V, Ng E, Chan C, Givoni B (2011) Outdoor thermal comfort study in a subtropical climate: a longitudinal study based in Hong Kong. International Journal of Biometeorology

Coceal O, Belcher SE (2004) A canopy model of mean winds through urban areas. Q J Roy Meteor Soc 130(599):1349–1372

Coceal O, Belcher SE (2005) Mean winds through an inhomogeneous urban canopy. Bound-Layer Meteorol 115(1):47–68

Finnigan JJ (ed) (1985) Turbulent transport in flexible plant canopies. in: The forest-atmosphere interaction. Reidel, Dordrecht

Finnigan JJ (2000) Turbulence in plant canopies. Annu Rev Fluid Mech 32:519–571

Frank J (2006) Recommendations of the COST action C14 on the use of CFD in predicting pedestrian wind environment. J Wind Eng 108:529–532

Gál T, Unger J (2009) Detection of ventilation paths using high-resolution roughness parameter mapping in a large urban area. Build Environ 44(1):198–206

Grell GA, Dudhia J, Staufeers DR (1994) A descrption of the fifth generation Penn State/NCAR Moesoscale Model (MM5). National Center for Atmospheric Research

Hall CT (2010) Predicting velocities and turbulent exchange in isolated street canyons and at a neighborhood scale. Massachusetts Insititute of Technology, Cambridge, US

Havens J, Walker H, Spicer T (1996). Wind tunnel study of air entrainment into two-dimensional dense gas plumes. Petroleum Environment Research Forum Project 93-16 Report

Hong Kong Development Bureau (HKDB) (2006) Technical Circular No. 1/06. Air ventilation assessment. Hong Kong Development Bureau, Hong Kong

Hong Kong Planning Department (HKPD) (2005) Feasibility study for establishment of air ventilation assessment system. Final report, The government of the Hong Kong Special Administrative Region

Kubota T, Miura M, Tominaga Y, Mochida A (2008) Wind tunnel tests on the relationship between building density and pedestrian-level wind velocity: development of guidelines for realizing acceptable wind environment in residential neighborhoods. Build Environ 43(10):1699–1708

Letzel MO, Krane M, Raasch S (2008) High resolution urban large-eddy simulation studies from street canyon to neighbourhood scale. Atmos Environ 42(38):8770–8784

MacDonald RW, Griffiths RF, Hall DJ (1998) An improved method for the estimation of surface roughness of obstacle arrays. Atmos Environ 32(11):1857–1864

Mochida A, Murakami S, Ojima T, Kim S, Ooka R, Sugiyama H (1997) CFD analysis of mesoscale climate in the Greater Tokyo area. J Wind Eng Ind Aerodyn 67–68:459–477

Murakami S (2004) Indoor/outdoor climate design by CFD based on the software platform. Int J Heat Fluid Flow 25:849–863

Murakami S (2006) Environmental design of outdoor climate based on CFD. Fluid Dyn Res 38(2–3):108–126

Ng E, Yuan C, Chen L, Ren C, Fung JCH (2011) Improving the wind environment in high-density cities by understanding urban morphology and surface roughness: a study in Hong Kong. Landscape Urban Plann 101(1):59–74

Petersen RL, Ratcliff MA (1989). Effect of homogeneous and heterogeneous surface roughness on heavier-than-air gas dispersion. American Petroleum Institute Publication No. 4491

Raupach MR, Shaw RH (1982) Averaging procedures for flow within vegetation canopies. Bound-Layer Meteorol 22(1):79–90

Raupach MR (1992) Drag and drag partition on rough surfaces. Boundary-Layer Meteorol 60:375
Schlichting H, Gersten K (2000) Boundary layer theory, 8th edn. Springer Science & Business Media, Germany
Skamarock WC, Klemp JB, Dudhia J, Gill DO, Barker DM, Wang W, Powers JG (2005) A description of the advanced research WRF version 2. NCAR Technical note. National Center for Atmospheric Research
Smargiassi A, Fournier M, Griot C, Baudouin Y, Kosatsky T (2008) Prediction of the indoor temperatures of an urban area with an in-time regression mapping approach. J Expo Sci Environ Epidemiol 18(3):282–288
Tominaga Y, Mochida A, Yoshie R, Kataoka H, Nozu T, Yoshikawa M, Shirasawa T (2008) AIJ guidelines for practical applications of CFD to pedestrian wind environment around buildings. J Wind Eng Ind Aerodyn 96:1749–1761
Wong MS, Nichol JE, To PH, Wang J (2010) A simple method for designation of urban ventilation corridors and its application to urban heat island analysis. Build Environ 45(8):1880–1889
Yim SHL, Fung JCH, Lau AKH, Kot SC (2007) Developing a high-resolution wind map for a complex terrain with a coupled MM5/CALMET system. J Geophys Res 112:D05106
Yim SHL, Fung JCH, Lau AKH, Kot SC (2009) Air ventilation impacts of the "wall effect" resulting from the alignment of high-rise buildings. Atmos Environ 43(32):4982–4994
Yuan C, Ng E (2014) Practical application of CFD on environmentally sensitive architectural design at high density cities: a case study in Hong Kong. Urban Clim 8:57–77
Yuan C, Ren C, Ng E (2014) GIS-based surface roughness evaluation in the urban planning system to improve the wind environment–a study in Wuhan, China. Urban Clim 10:585–593
Zhu G, Arya SP, Snyder WH (1998) An experimental study of the flow structure within a dense gas plume. J Hazard Mater 62(2):161–186

Part III
Building Scale Wind Environment

Chapter 5
Building Porosity for Better Urban Ventilation in High-Density Cities

5.1 Introduction

5.1.1 Background

Nowadays, cities are homes to over half of the world's population, and the city population is expected to reach 4.96 billion by 2030 (United Nations Population Division 2006). Urbanization improves living quality but also spontaneously increases the demand on natural resources (World Wildlife Fund (WWF) 2010). To address such urbanization issue, high-density living that enables cities to utilize resources more efficiently by decreasing traffic cost and other energy usage is a viable alternative (Betanzo 2007). However, high-density living may lead to congested urban conditions, which in term are associated with serious environmental issues, such as poor outdoor natural ventilation (Ng 2009a). Taking Hong Kong as an example, the extremely rapid and successful growth of Hong Kong in the last century had led to the increase of high-rise compact buildings and deep street canyons (Fig. 5.1), which significantly block airflow in the urban canopy layer (Fig. 5.2).

Stagnant air at urban areas is associated with both outdoor urban thermal discomfort and increased air pollution. Outdoor thermal comfort under typical summer conditions requires 1.6 m/s wind speed; however, a decrease in wind speed from 1.0 to 0.3 m/s during the summer could result in 1.9 °C increase in temperature (Cheng et al. 2011). Wind data from urban observatory station of Hong Kong Observatory (HKO) indicates that the mean wind speed at 20 m above the ground level in a urban area (i.e., Tseung Kwan O) has decreased by about 40%, from 2.5 to 1.5 m/s over the past

---

Originally published in Chao Yuan and Edward Ng, 2012. Building porosity for better urban ventilation in high-density cities–A computational parametric study. Building and Environment, 50, pp.176–189, © Elsevier, https://doi.org/10.1016/j.buildenv.2011.10.023.

C. Yuan, *Urban Wind Environment*, SpringerBriefs in Architectural Design and Technology, https://doi.org/10.1007/978-981-10-5451-8_5

decade (Hong Kong Planning Department (HKPD) 2005). Due to poor dispersion, frequent occurrences of high concentrations of pollutants, such as NO_2 and respirable particles (RSP), in urban areas like Mong Kok and Causeway Bay have been reported by the Hong Kong Environmental Protection Department (Yim et al. 2009b).

The 2003 outbreak of the Severe Acute Respiratory Syndrome (SARS) epidemic in Hong Kong had brought attention to how environmental factors (i.e., air ventilation and dispersion in buildings) played an important role in the transmission of SARS and other viruses. Since the outbreak, the Hong Kong Special Administrative Region (HKSAR) Government strived to improve the local wind environment for better urban living quality. Relevant studies, policies, and technical guidelines, such as Air Ventilation Assessment (AVA) and Sustainable Building Design (SBD) Guidelines, have been conducted and applied on the urban planning and design process (Ng 2009b). The Hong Kong Planning Department (HKPD) initiated the "Feasibility study for establishment of Air Ventilation Assessment system" in 2003 to answer the fundamental question of "*how to design and plan our city fabric for better natural air ventilation?*" (HKPD 2005). In that study, the current air ventilation issues in Hong Kong were provided, and the corresponding qualitative urban design methodologies and guidelines were given in order to create an acceptable urban wind environment. Unlike AVA, the SBD Guidelines (under APP-152 of Buildings Department HKSAR) are intended for building scale, providing the quantitative requirements for three building design elements, namely, building separation, building set back, and site coverage of greenery. The SBD Guidelines (1) aim to mitigate the negative effects of new building development on the existing surrounding wind environment, (2) enhance the environmental quality of living spaces in Hong Kong, and (3) enable architects to mitigate the undesirable effect of new buildings without needing CFD simulation and wind tunnel experiments by modifying simple design indexes (e.g., building length, distance to the adjunct streets or site boundary, width of the building gap, and size of wind permeability) from the Guidelines.

The computational parametric study described in this chapter aims to quantify the SBD Guidelines by evaluating the effect of mitigation strategies on improving urban wind environment. Scientific evidence-based understandings are provided in order to choose appropriate design strategies that can efficiently improve the pedestrian-level natural ventilation while preserving land use efficacy. As shown in Fig. 5.3, Mong

Fig. 5.1 View of the Kowloon Peninsula and Hong Kong Island in the 1900s and today (Waller 2008; Moss 2008)

Fig. 5.2 High-rise compact building blocks and deep street canyons in Hong Kong (Wolf 2009)

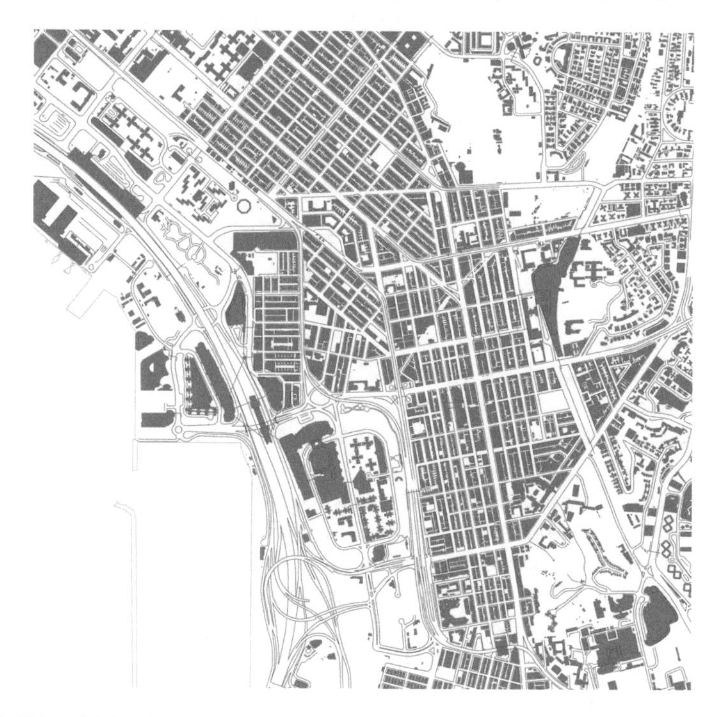

Fig. 5.3 Urban fabric of Mong Kok: the grid plan

Kok, one of metropolitan areas in Hong Kong, is chosen as the case study to represent regular street grid planning, which is widely applied in mega cities planning, such as Manhattan, and Philadelphia in the U.S.

5.2 Literature Review

5.2.1 *Outline of CFD Numerical Methods for Neutral Turbulence Flows*

The rapid development of CFD application in environmental design in the past three decades is remarkable. Not only is CFD a well-established environmental research tool to enhance our predictive power, it is also a recognized design tool for urban planners and designers (Murakami 2006). With CFD, highly unsteady and three-dimensional turbulent flows in the atmospheric boundary can be simulated. In the following sections, three main CFD methods, which are Direct Numerical Simulation (DNS), Large Eddy Simulation (LES), and Reynolds-averaged Navier–Stokes (RANS) are discussed.

Compared with laminar flows, the fluctuation of turbulent flows is on a broad range of spatial and temporal scales (Ferziger and Peric 2002; Murakami 2006). In DNS, all spatial and temporal scales of the turbulence motions are directly computed by Navier–Stokes equations without any approximations by turbulence models (Ferziger and Peric 2002). Because of this, DNS is the most accurate among all of the simulation methods. However, using the DNS model to totally reproduce turbulence is extremely expensive. On the other hand, the LES method separates the turbulence flow into large and small scales, and only focuses on the "large eddy" that is larger than grid size. Small eddies, such as eddies in Kolmogorov scales, are eliminated from the solution to reduce computational cost. The unresolved scales of turbulence flows are approximated by the sub-filter-scale turbulence models (Meneveau and Katz 2000; Kleissl and Parlange 2004). As a result, LES is computationally cheaper than DNS, and has been used as a research tool in a wide range of engineering applications to simulate high Reynolds flows, such as simulations of the neutral atmospheric boundary layer (Letzel et al. 2008).

Unlike the DNS and LES methods, which are time-dependent approaches, the RANS method does not directly compute any turbulence by Navier–Stokes equations, but rather approximates the turbulence flows by decomposing solution variables in DNS and LES into the time-averaged and fluctuating components:

$$f = <f> + f' \tag{5.1}$$

where f is the instantaneous value of variables, such as velocity, in the Navier–Stokes equations, $<f>$ is the mean value, and f' is the fluctuating value. The turbulence models are to approximate the effects of f'. Therefore, compared with DNS and LES, RANS is a simplified engineering approximation that has been broadly applied as a design tool.

The computational parametric study described in this chapter applies the RANS method to simulate the wind velocity in the neutral condition. Reynolds-averaged equations are yielded by substituting Eq. (5.1) to the Navier–Stokes equations. The incompressible flow equations with constant density (ρ), i.e., airflow in this study, can be written as follows:

$$\text{Continuity equation}: \quad \frac{\partial <u_i>}{\partial x_i} = 0 \tag{5.2}$$

$$\text{Momentum equation}: \quad \frac{D<u_i>}{Dt} = -\frac{1}{\rho} \cdot \frac{\partial <P>}{\partial x_i} + \frac{\partial}{\partial x_j}\left(\upsilon \frac{\partial <u_i>}{\partial x_j} - <u_i' u_j'>\right), \tag{5.3}$$

where x_i ($i = x, y, z$) are the coordinates, u_i are velocity vectors, t is time, v is dynamic viscosity, and p is pressure. Equations (5.2) and (5.3) are similar with the Navier–Stokes equations, except that all variables are time averaged, and the Reynolds stresses, $-<u_i' u_j'>$, must be approximated by turbulence models to close Equations.

According to the various approaches of dealing with Reynolds stresses, $-<u_i' u_j'>$, the RANS model can be classified into two types: (a) Boussinesq Approach–turbulence viscosity ($\kappa - \varepsilon$ and $\kappa - \omega$ turbulence models) and (b) Reynolds Stress Models (RSM). The RSM turbulence model, as an anisotropic model, considers the effects of Reynolds stress by using six equations to compute $-<u_i' u_j'>$ directly. Compared to the $\kappa - \varepsilon$ and $\kappa - \omega$ turbulence models, where ε (turbulence dissipation rate) or ω (the specific dissipate rate) is assumed isotropic, RSM is more computationally costly. However, RSM is more accurate than the $\kappa - \omega$ realized model in flows with strong anisotropic effects (i.e., swirling flows) (Kuznik et al. 2007). On the other hand, the accuracy of RSM is similar to the models by Boussinesq Approach in normal flows (Wong 2004).

Current commercial CFD codes provide both turbulence viscosity model and RSM models. In our computational parametric study, the $\kappa - \omega$ SST (Shear Stress Transport) models were applied. The $\kappa - \omega$ SST model is a combination of the standard $\kappa - \omega$ model and $\kappa - \varepsilon$ model. Walls are the main source of turbulence; therefore, an accurate near-wall treatment is critical for turbulence models to be useful (Fluent 2006). While the standard $\kappa - \omega$ model, as a near-wall model, is more accurate than the $\kappa - \varepsilon$ models in the near-wall layers (Menter et al. 2003), it cannot replace the $\kappa - \varepsilon$ models in the simulation of the outer part of the near-wall region (Menter et al. 2003). The $\kappa - \omega$ SST model uses the standard $\kappa - \omega$ model for the inner part and gradually changes to the $\kappa - \varepsilon$ model for the outer part (Menter et al. 2003; Fluent 2006).

5.3 Validation

Cross-comparisons of wind data from RANS ($\kappa - \varepsilon$ models), DES, LES, and the wind tunnel experiments were conducted by a working group from the Architectural Institute of Japan (AIJ) to calibrate the CFD simulation results (Tominaga et al.

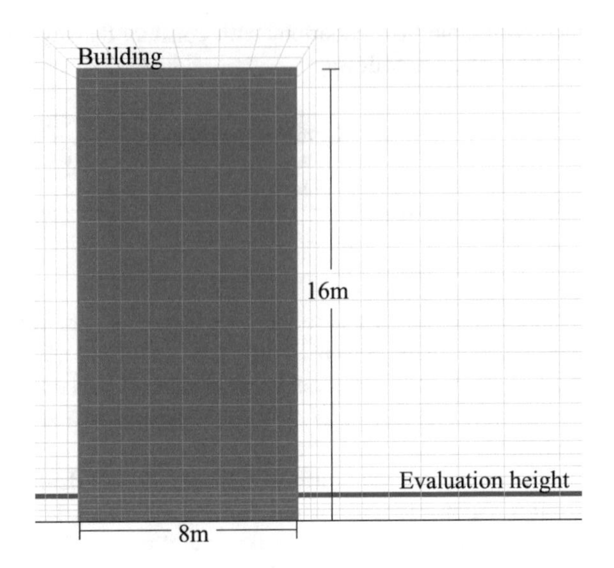

Fig. 5.4 Grid structure and resolution. The maximum grid size ratio is 1.2 and four mesh layers are arranged below the evaluation height

2008; Mochida et al. 2002). In this chapter, the SST $\kappa - \omega$ model was validated by comparing the simulation results with the wind tunnel data from AIJ. Menter et al. (2003) compared the simulation results obtained from the $\kappa - \omega$ SST and DES SST models. The data for comparison was a set of vertical mean wind velocity profiles across a single cubic block. The result suggests that while the $\kappa - \omega$ SST model can predict airflow near the building, it fails to reproduce the recovery downstream in the separation zone that is far from the building (Menter et al. 2003). Since our study mainly focuses on the pedestrian-level wind environment, both the pedestrian-level wind speed distribution and vertical wind velocity profiles are compared with the data derived from the wind tunnel experiment.

CFD simulation settings for the validation and parametric study follow AIJ guidelines for urban pedestrian wind environment modeling. The size of computational domain and building block in the validation study are $500 \times 500 \times 60$ m and $8 \times 8 \times 16$ m ($W \times L \times H$), respectively. As shown in Fig. 5.4, adaptive meshing method is applied to reduce computational cost, and accurately predicted wind speed at the areas of interest. The finer scale grids are arranged at the areas around the building and close to the ground (Fig. 5.4). To comply with the AIJ guidelines, four layers (layer height: 0.5 m) are arranged below the evaluation height (2.25 m above the ground). The maximum grid size ratio is set to 1.2, as shown in Fig. 5.4.

In terms of input boundary condition, the vertical wind speed profile is set to

$$U_{(h)} = U_{\text{met}} \cdot \left(\frac{h}{d_{\text{met}}}\right)^{\alpha}, \tag{5.4}$$

where α is the surface roughness factor $\alpha = 0.2$ U_{met} is the wind speed at the top of the domain ($U_{\text{met}} = 6.751$ m/s), d_{met} is the domain thickness ($d_{\text{met}} = 60$ m), and h

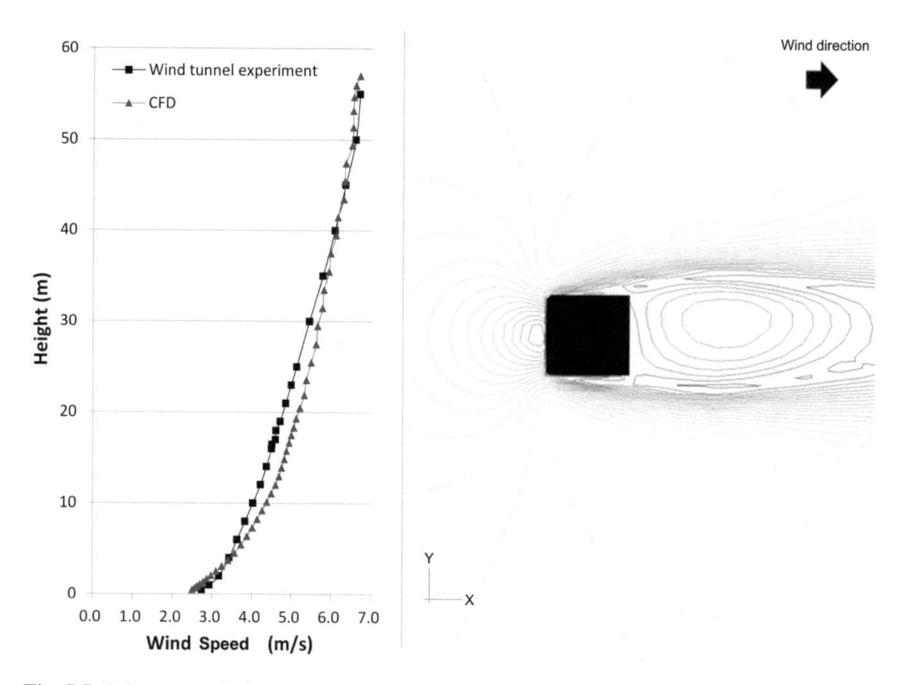

Fig. 5.5 Inlet mean wind speed profiles in the CFD simulation and wind tunnel experiment, and the simulation result (the contour of wind speed at 2.25 m above the ground)

is the height. To compare with data from the wind tunnel experiment, the inlet wind profile in CFD simulation is set similarly as possible to the one from the wind tunnel experiment, as shown in Fig. 5.5.

The contour of wind speed at 2.25 m above the ground is presented in Fig. 5.5. Figure 5.6 shows the position of the test points and cross-comparing result (R^2 = 0.853), which indicates that the $\kappa - \omega$ SST model can accurately predict the pedestrian-level airflow within urban context. To further validate the simulation results, the vertical profiles of the mean wind velocities from CFD simulation and wind tunnel experiment at position 1 and position 2 (depicted in Fig. 5.6) are also presented in Fig. 5.7. The results are consistent with the experiment results produced by Menter et al. (2003).

5.4 Computational Parametric Study

In this section, a parametric approach is applied to obtain more general understandings on wind performance with particular building typologies, and to identify the effects of individual design parameters on natural ventilation performance more easily than do conventional studies (Mochida et al. 1997; Kondo et al. 2006; Ashie et al. 2009). Specifically, three groups with 27 testing scenarios are set up in three

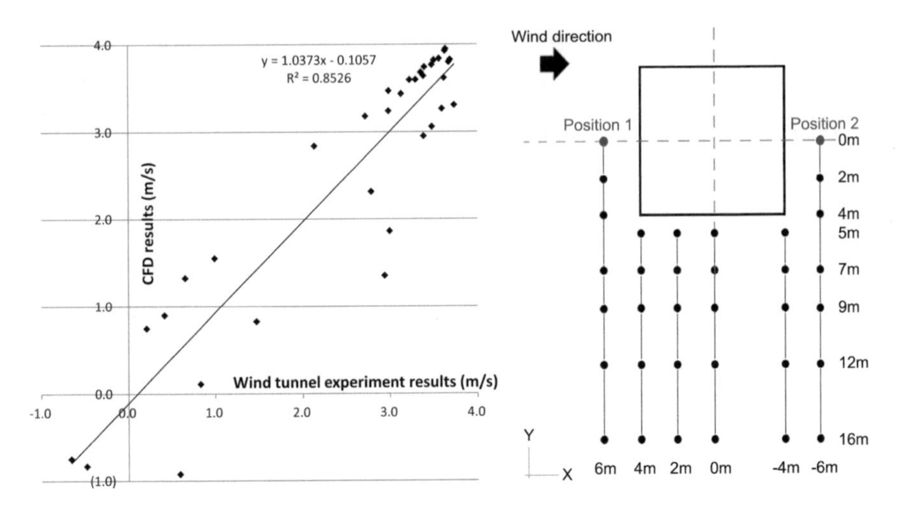

Fig. 5.6 Test points and the linear correlation of wind speeds from the CFD simulation and wind tunnel experiment (significant level: 0.95)

input wind directions, and are simulated by the $\kappa - \omega$ SST model. The following section will introduce the building model design, computational modeling of the testing scenario, and boundary conditions in this computational parametric study.

5.4.1 Parametric Models

To evaluate the effect of urbanization on urban natural ventilation and to predict future conditions, three design scenarios (i.e., Cases 1903, 1, and 2) are designed based on the urban morphologies of the past, present, and future Mong Kok, a metropolitan area that has significantly changed overtime. Compared with Mong Kok in 1903, the population density in present day Mong Kok is extremely high, reaching 130,000/km^{-2} (Wikipedia 2011). Given the current planning trend, the population density of this downtown area in the future will be much higher. Case 1903 refers to the past Mong Kok characterized by three or four-floor buildings with a sloping roof. In a simplified parametric model, the sloping roof is replaced by a flat floor and 20 m building height. Case 1 refers to the present day Mong Kok in which the height of buildings and podiums is on average 45 and 15 m, respectively (Ng et al. 2011). The site area is not fully occupied, with narrow gap in between. Case 2 refers to future Mong Kok in which the height of buildings on average is estimated to increase to 90 m. The site area will be fully occupied in response to the increasing urban land use.

Six design options with various building typologies are tested to see the effects of different building typologies on pedestrian-level natural ventilation performance in order to identify efficient mitigation strategies. As shown and summarized in Fig. 5.8, mitigation strategies include: (a) setting back buildings, (b) separating the

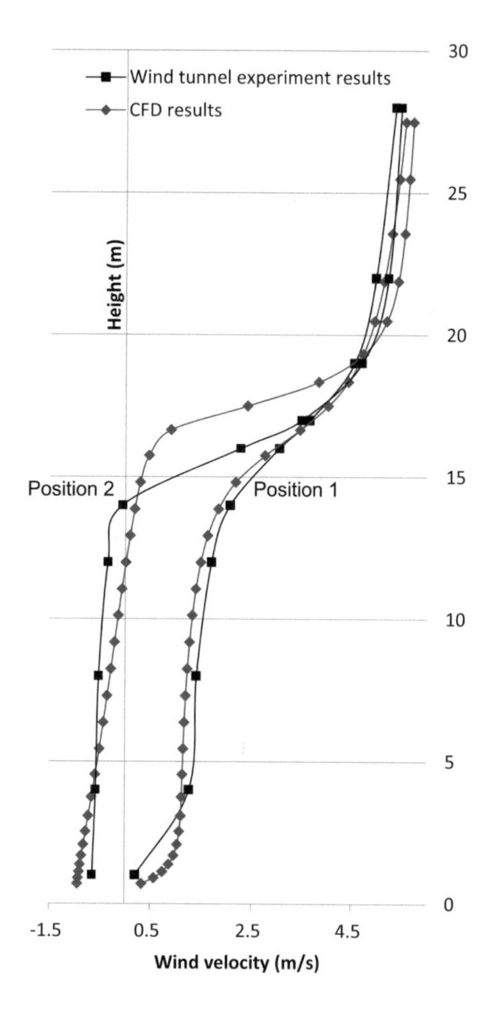

Fig. 5.7 Vertical profiles of the mean wind velocities from the CFD simulation and wind tunnel experiment. Locations of the vertical lines: Position 1 (windward) and 2 (leeward)

long buildings, (c) stepping the podium, (d) opening the permeability of towers and podiums, and (e) creating a building void between towers and podiums. In Cases 3–7, the building volumes are similar to the one in Case 2 to keep the land use density constant.

Case 3 is similar to Case 2, except that the building set back is 15 m along the street in Case 3. Case 3 shows that setting back buildings is relatively easy to be applied in urban planning. Building separations along the prevailing wind direction are applied in Case 4. The decreased land use efficiency resulting from the building separations is compensated by higher towers (123 m). Case 5 is the combination of Cases 3 and 4. The stepped podium and building void in Case 6 are popular design strategies in Hong Kong, which are favorable for natural outdoor ventilation and greening. Air passages in high and wide buildings are applied in Case 7. This strategy is currently widely used in Hong Kong to mitigate the undesirable effects on the leeward wind environment.

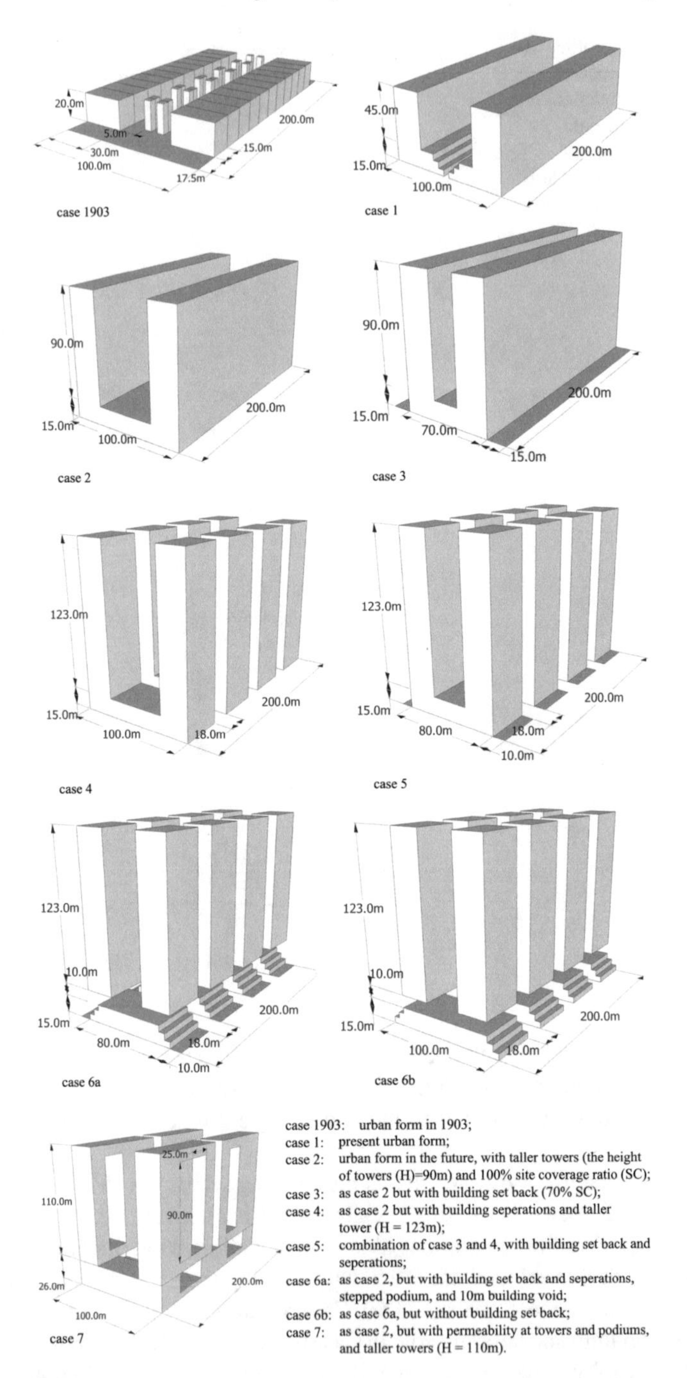

Fig. 5.8 Summary of nine testing models which are designed to test the effects of different building morphologies on natural ventilation performance

5.4.2 Computational Modeling

A computational modeling for neighborhood scale is used in the current parametric study. Several existing parametric studies for wind evaluation and prediction have focused on the airflow around only one or two simple and generic buildings (Stathopoulos and Storms 1986; To and Lam 1995; Blocken et al. 2007). However, the turbulence kinetic energy produced by the urban context is difficult to reproduce in the inlet boundary condition. The testing scenario modeled in the urban scale can approximately include the effect of urban context on airflow to evaluate and predict the natural ventilation performance in the area of interest more accurately.

In the current study, 27 testing scenarios are simulated in the neutral condition by the $\kappa - \omega$ SST model in three wind directions (0°, 45°, and 90°). As shown in Fig. 5.9, each testing scenario is made up of a corresponding model array (6×10) and two rows of surrounding randomized buildings. The same regular street grid is applied in all model arrays; the width of the street canyon is 20 m. To consider the effects of urban context, only the wind data measured at the target area (Fig. 5.9) are used in the numerical analysis.

The computational domain size is $3.2 \times 3 \times 0.45$ km ($W \times L \times H$), as shown in Fig. 5.9. The grid point numbers are 5.0–6.8 million on a case-by-case basis. Similar to the meshes in the validation study, the finer meshes are arranged at the areas around the buildings and close to the ground. To comply with the AIJ guideline, the maximum grid size ratio is set to 1.2, and the three grid layers (layer height: 1 m) are arranged below the evaluation level. This means that only the wind data at 3.5 m above the ground are collected and analyzed.

To set the inlet boundary condition for Mong Kok area by Eq. (5.4), the site-specific annual wind rose data measured at a 450 m height are obtained from the fifth-generation NCAR/PSU mesoscale model (MM5) (Yim et al. 2009a). As shown in Fig. 5.10, wind from east and northeast have higher and the highest probability of occurrence, respectively. Therefore, input prevailing wind direction in the simulation is set to 90°. Simulations in the input wind directions of 0° and 45° (Fig. 5.10) are also included to describe comprehensively the natural ventilation performance in different wind directions. Wind speed at the top of the domain is set to $U_{\text{met}} = 11$ m/s, which is the mean prevailing wind speed at a 450 m height in the summer. Owing to the high urban surface roughness, surface roughness factor (α) is set to 0.35 in this computational parametric study. The outlet wind profile is set to be the same as the inlet wind profile.

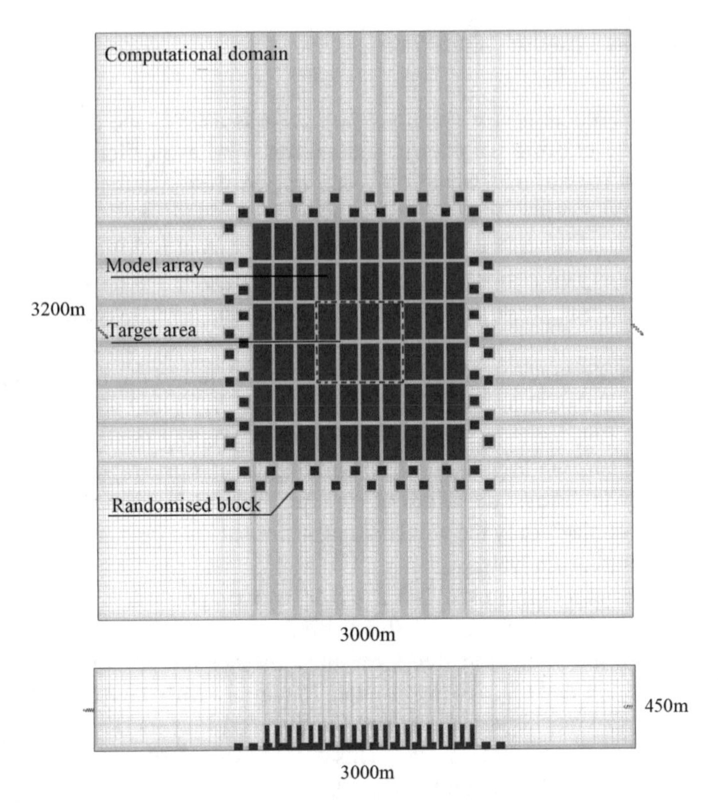

Fig. 5.9 Computational domain and grids for the parametric study. The model array, randomized block, and the target area are presented

5.5 Modeling Results and Analysis

5.5.1 Wind Speed Classification

A wind speed classification based on outdoor thermal comfort is derived to correlate the building design options with outdoor thermal comfort. Different from the classification by Beaufort (cited in Cheng and Ng 2006) and Murakami and Deguchi (1981), in which wind is a kind of "nuisance", the current parametric study considers wind as a benefit. Therefore, this classification is based on the outdoor thermal comfort and not on the wind force on pedestrians or buildings. Ng et al. (2008) conducted a survey to use physiological equivalent temperature (PET) to obtain the pedestrian-level wind speed threshold values, especially for outdoor thermal comfort in Hong Kong (Hoppe 1999). This survey showed that a wind speed of 0.6–1.3 m/s was required to achieve neutral thermal sensation (neutral PET: 28.1 °C) in a typical summer day with ambient temperature of 27.9 °C and relative humidity of 80%. In another survey on thermal comfort in Hong Kong (Cheng et al. 2011), the wind speed settings

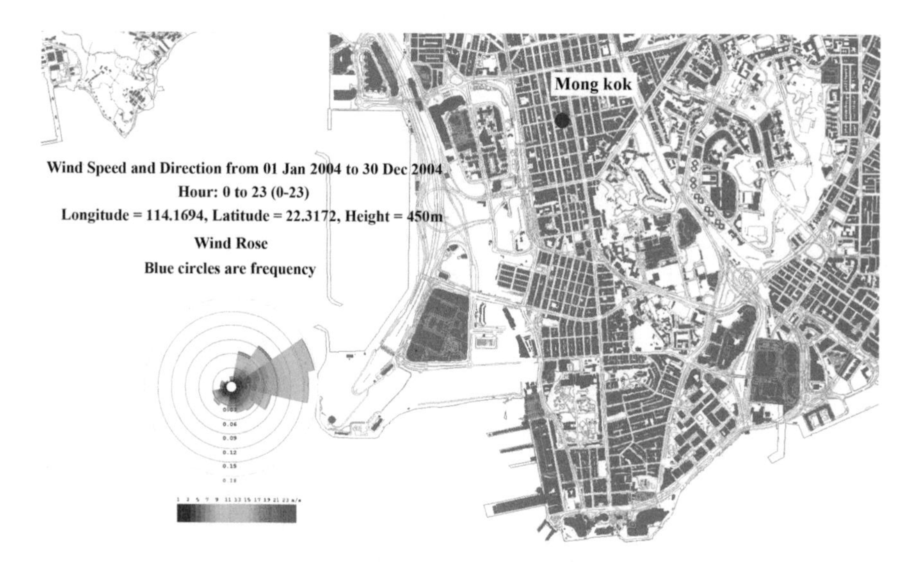

Fig. 5.10 Site-specific annual wind rose of Mong Kok

with and without wind break were 0.3 m and 1.0 m, respectively, which equaled to a 1.9 °C difference in ambient temperature. Based on this literature review, the classification of pedestrian-level wind speed (u) is assigned as the following: Class 1: $u < 0.3$ m/s for "stagnant," Class 2: 0.6 m/s $> u \geq 0.3$ m/s for "poor," Class 3: 1.0 m/s $> u \geq 0.6$ m/s for "low," Class 4: 1.3 m/s $> u \geq 1.0$ m/s for "satisfactory," and Class 5: $u \geq 1.3$ m/s for "good" pedestrian-level natural ventilations in street canyons, respectively.

5.5.2 Impact of Input Wind Directions

The contour colored by mean wind speed in Case 1 is presented in Figs. 5.11 and 5.12. The accelerated wind speed is found at street canyons that are parallel with the input wind direction (wind direction 90°, Fig. 5.11). However, other than these areas, wind speed is significantly reduced in deep street canyons perpendicular to the wind direction. When the input wind direction is changed from 90° (Fig. 5.11) to 45° (Fig. 5.12, right), air can enter into the streets from both directions (Fig. 5.12), resulting in the decrease in areas with stagnant air. On the other hand, air can enter more deeply into the streets in the wind direction 0° (Fig. 5.12, left) than 90° (Fig. 5.11), due to the smaller frontal area (please refers to Chap. 2 for more details and definition of frontal area). The above discussion is consistent with the statement in the AVA study saying "*An array of main streets, wide main avenues and/or breezeways should be aligned in parallel, or up to 30 degrees to the prevailing wind direction, in order to maximize the penetration of prevailing wind through the district.*" (HKPD 2005).

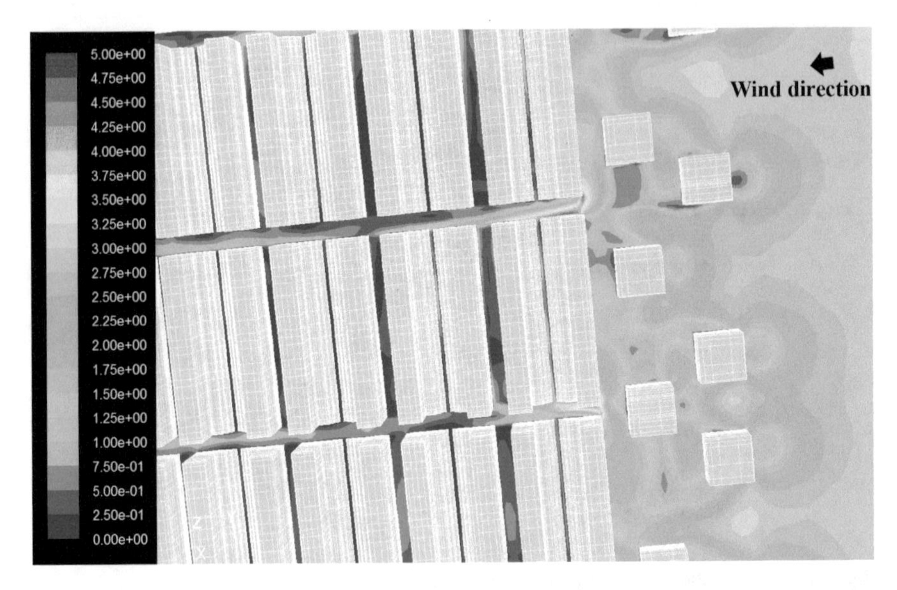

Fig. 5.11 Contour of wind speed in the wind direction of 90° (prevailing wind direction) at 3.5 m above the ground (Case 1). (For interpretation of the references to color in the text, the reader is referred to the electronic version of this book)

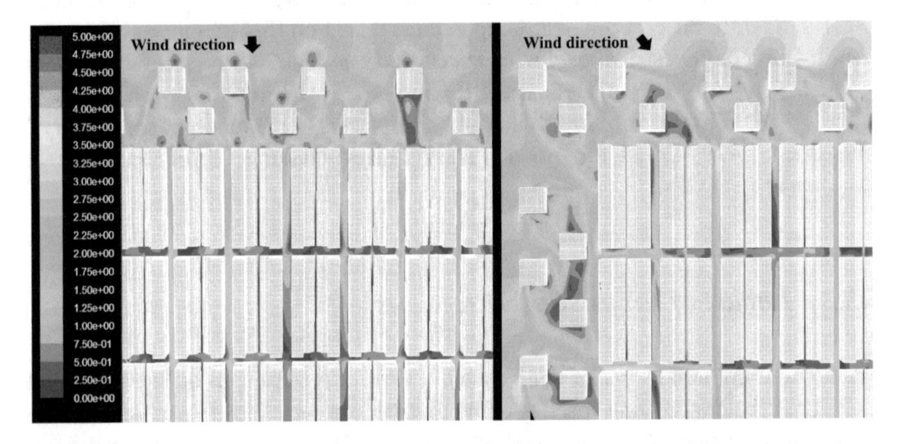

Fig. 5.12 Contours of wind speed in the wind direction of 0° (right) and 45° (left) at 3.5 m above the ground (Case 1). (For interpretation of the references to color in the text, the reader is referred to the electronic version of this book)

5.5.3 Impact of Building Typologies

A cross-comparison of natural ventilation performance in the prevailing wind direction of 90° is summarized in Figs. 5.13 and 5.14, and Table 5.1. Wind data are statistically analyzed based on the relative frequency of the wind speed classifica-

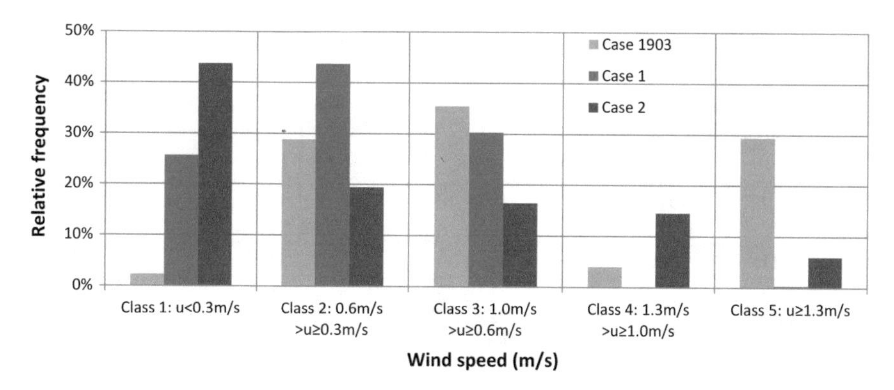

Fig. 5.13 Relative frequencies of pedestrian-level mean wind speed in Cases 1903, 1, and 2 (Input wind direction: 90°)

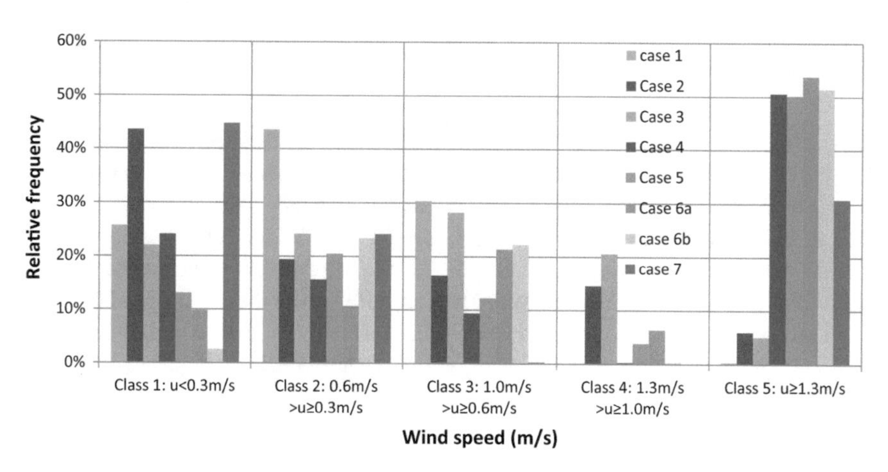

Fig. 5.14 Relative frequencies of the pedestrian-level wind speed in Cases 1, 2, 3, 4, 5, 6a, 6b, and 7 (Input wind direction: 90°)

tion. About 300 wind speed data in every test scenario are collected at the centerlines of four streets in the target area (Fig. 5.9). Figure 5.13 shows the evaluation of natural ventilation performance of the past (Case 1903), present (Case 1), and future (Case 2) Mong Kok. For area with very poor natural ventilation (Class 1: $u < 0.3$ m/s), Case 1903 accounts for the lowest frequency (2.29%), whereas Cases 1 and 2 account for 25.67 and 43.67%, respectively. This is consistent with the rapid urbanization trend in the Mong Kok area in the last century.

The effects of design options with various mitigation strategies (Cases 3–7) on the urban natural ventilation performance are presented in Fig. 5.14. First, building setbacks (Case 3) and building separations (Case 4) are helpful in improving pedestrian-level natural ventilation. Compared with Case 2, the relative frequencies of very poor ventilation (Class 1) in Cases 3 and 4 reduce to 22.02 and 24.12%, respectively. However, the wind performances in Cases 3 and 4 still resemble that in

Table 5.1 Summary of the relative frequencies of pedestrian-level wind speed in nine testing models (input wind direction: 90°)

	$u < 0.3$ m/s	0.6 m/s $> u \geq 0.3$ m/s	1.0 m/s $> u \geq 0.6$ m/s	1.3 m/s $> u \geq 1.0$ m/s	$u \geq 1.3$ m/s
	Stagnant (%)	Poor (%)	Low (%)	Satisfactory (%)	Good (%)
Case 1903	2.29	28.86	35.43	4.00	29.43
Case 1	25.67	43.67	30.33	0.00	0.33
Case 2	43.66	19.40	16.42	14.55	5.97
Case 3	22.02	24.19	28.16	20.58	5.05
Case 4	24.12	15.59	9.41	0.29	50.59
Case 5	13.06	20.47	12.17	3.86	50.45
Case 6a	9.83	10.69	21.39	4.34	53.76
Case 6b	2.60	23.41	22.25	0.29	51.45
Case 7	44.78	24.18	0.27	0.00	30.77

Case 1 (the current urban wind environment), and thus are deemed to be suboptimal. Second, the relative frequency of Class 1 is decreased further to 13.06% in Case 5, in which both building setbacks and separations are adapted. Third, stepped podiums and 10 m building voids are applied in Cases 6a and 6b. The relative frequency of Class 1 in Case 6a and Case 6b is 9.83 and 2.60%, respectively, which indicate good natural ventilation performance similar to Case 1903, even though the land use density in Case 6 is higher than that in Case 1903. Last but not least, air passages in the tower and podium are applied in Case 7, with very high relative frequency (44.78%) in Class 1 remaining.

Furthermore, two similar analyses are conducted for the cases in the input wind directions of 0° and 45°. First, as shown in Figs. 5.15 and 5.16, the relative frequencies of good natural ventilation performance reach to 80%, which indicates that the modification of street grid direction is helpful in improving natural ventilation performance. Second, the natural ventilation performance becomes less sensitivity to the changes in building morphologies in these two input wind directions, especially in the 45° input wind direction.

5.5.4 Comparison of the Vertical Wind Profiles

The vertical profiles of the mean wind velocity are collected at a point on the centerline of the street across the prevailing wind direction, and they are plotted in Fig. 5.17 to help better understand the above analysis results. Figure 5.17a shows the profiles from 0 to 400 m, indicating that building height significantly affects wind profiles, especially at $z_d + z_0$ values. The height of zero velocity in the wind profiles is close to but slightly lower than the mean height of buildings. This result is consistent with

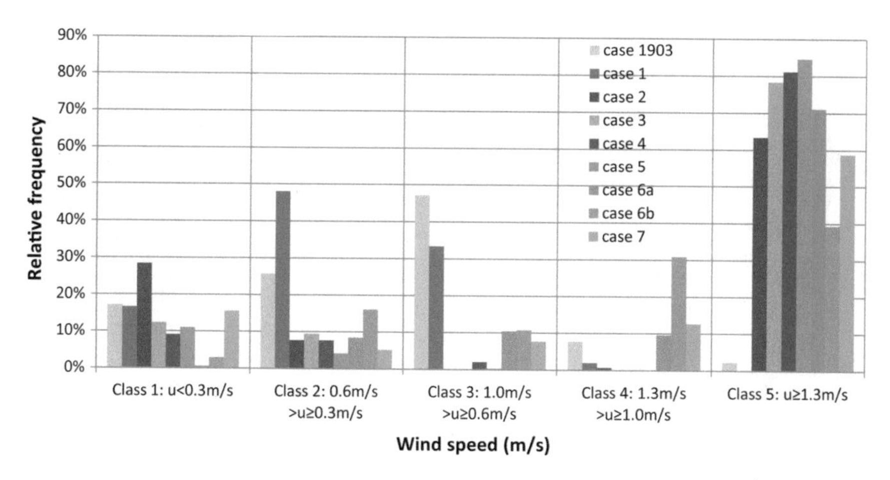

Fig. 5.15 Relative frequencies of the pedestrian-level wind speed in Cases 1903, 1, 2, 3, 4, 5, 6a, 6b, and 7 (Input wind direction: 0°)

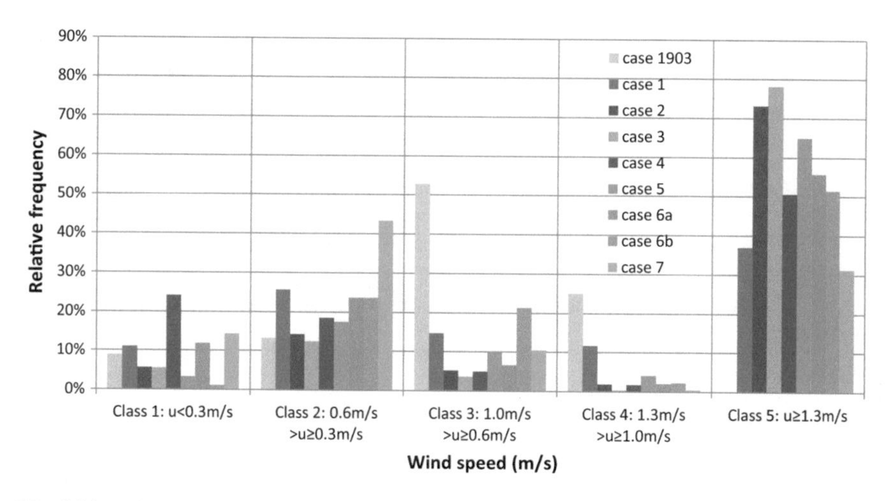

Fig. 5.16 Relative frequencies of the pedestrian-level wind speed in Case 1903, 1, 2, 3, 4, 5, 6a, 6b, and 7 (Input wind direction: 45°)

that of Oke (2006) and Lawson (2001). The profiles from 0 to 50 m are enlarged in Fig. 5.17b to show airflow that is closer to the ground. Our results indicate that the wind environment at the pedestrian level is not affected by building height, but rather is significantly influenced by building morphology in the podium layer. The results agree with the frontal area density study in the urban scale by Ng et al. (2011).

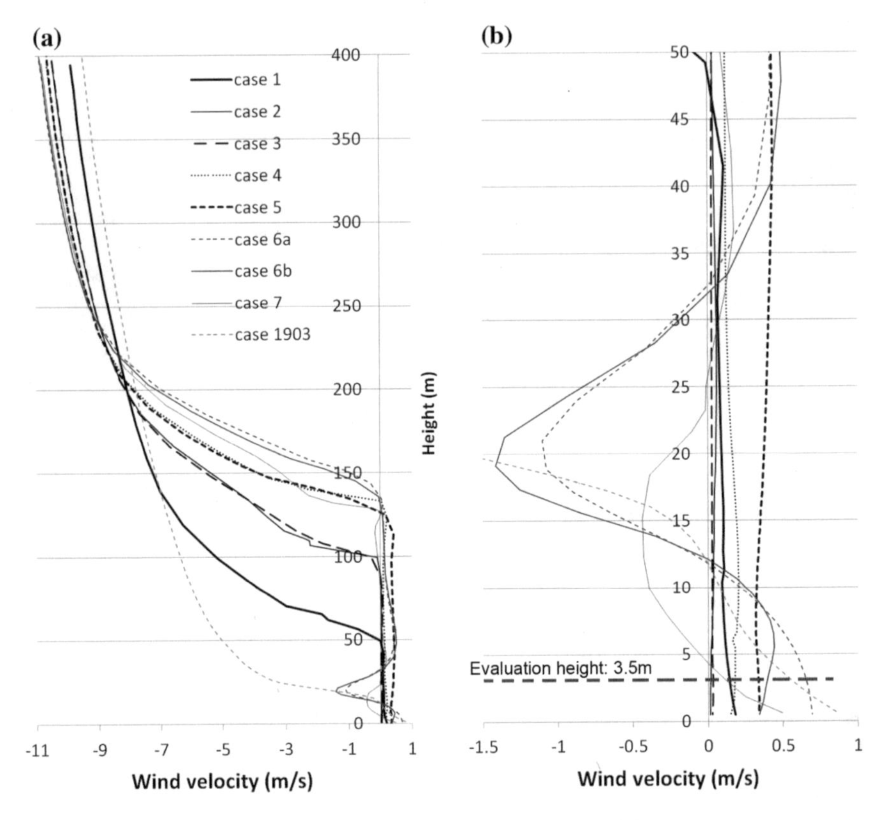

Fig. 5.17 Mean wind velocity profiles measured at the centerline of the street canyon across the wind direction

5.6 Conclusion

This chapter presents a computational parametric approach for evaluating the effects of different urban morphologies on the pedestrian-level natural ventilation environment through CFD simulation. Different simulation methods and turbulence models are compared based on the literature review. The $\kappa - \omega$ SST turbulence model is applied in this study. Results of the comparison between CFD simulation results and wind tunnel data suggest that the $\kappa - \omega$ SST turbulence model is accurate enough to simulate the turbulent flows caused by shape-edged buildings. Unlike the wind field studies in real urban areas, our study designs a number of parametric models to test the different modification effects on pedestrian-level air flow in street canyons. Compared with other parametric studies in which only one or two buildings are included, this study employs computational modeling in the urban scale to include the effects of the urban context. This study derives a wind speed classification based on the outdoor thermal comfort, as well as uses the relative frequency of the pedestrian-level wind speed and the vertical profiles of the mean wind velocity to evaluate the natural ventilation performance in street canyons. Cross-comparison results suggest that the sensitivity of the wind performance is not the same across different modifications of building morphology. Relevant planning and design strategies are provided based on the findings of this study as follows:

(1) Street grid orientation in grid planning is a significant parameter in urban natural ventilation performance. Main streets should be arranged along the prevailing wind direction. Efficiency of the design strategies also depends on the prevailing wind direction, as demonstrated in the results in Sect. 5.3.
(2) The mean building height determines the $z_0 + z_d$ value in high-density urban areas, while urban ventilation performance from the pedestrian perspective mostly depends on the pedestrian-level building porosity.
(3) Overall, decreasing the site coverage ratio increases the pedestrian-level natural ventilation performance. Comparison of Cases 3, 4, and 5 provides a more detailed understanding. The wind profiles in Cases 3 and 4 suggest that building setbacks along the street across the prevailing wind direction (Case 3) are less useful in high-density urban areas than building separations along the prevailing wind direction (Case 4). In contrast, the wind profiles in Case 5 (Fig. 5.8) suggest improvement of the pedestrian-level wind speed on the leeward outdoor space if building separation is incorporated with building setback.
(4) Wind permeability in the podium layer is very useful in leading airflow into deep street canyons. The building void between towers and podiums in Case 6 significantly accelerates turbulent flows in the podium layer, and the stepped podiums lead airflow into the pedestrian level. Airflow across the building void (10 m) is numerically presented by the red lines in Fig. 5.8. The mean wind speed on the leeward outdoor space can be improved to recreate the condition in Case 1903.
(5) Air passages should be arranged as close to the ground level as possible. Poor wind performance in Case 7 indicates that wind permeability in towers does not improve the pedestrian-level wind environment. Furthermore, compared with the profiles of Case 6, wind passage in the podium layer should incorporate the stepped podium to benefit the pedestrian-level wind environment, or the openings on the facade should be opened from the ground.
(6) Building setback, separation, and building permeability are useful in improving the pedestrian-level wind environment. However, the efficiency levels of these strategies differ. The consideration of urban area as a whole is important as natural ventilation performance in urban area results from the integral effects of buildings. Thus, air paths in these areas can be efficiently established and organized by applying different strategies to improve building porosity. Combining strategies (i.e., urban planning + building design) is recommended for the doubled benefits and efficiency over any single strategy. Planners and architects should choose appropriate strategies based on the actual design requirements and the insights from this Chapter.

(7) The current study gives scientific insights into the urban ventilation section of the Hong Kong Planning Standards and Guidelines (HKPSG) (HKPD 2011), as the Guidelines stress the importance to design urban breezeway and air path with nonbuilding areas and building setbacks from streets. Proper street alignment and sufficient open spaces that can be interlinked are equally important, as well as integration of the urban planning efforts into building-level designs. Building gaps, separations, and porosity close to the pedestrian levels are extremely useful in improving air space for urban air ventilation circulation.

(8) Increasing urban density has gained attention in recent worldwide discussions. Traditional wisdom attributes the decline of urban air ventilation to increasing urban density because of high urban frontal area density (FAD). While this understanding is generally correct, balancing the need to reduce land resources demand with designing more compact cities is also useful. The current study provides important insights on fine-tuning and designing urban morphologies and building forms that optimize urban air ventilation. Such optimization process can further incorporate the bio-meteorological needs of air ventilation in different climatic zones (Cheng et al. 2011).

(9) Evidence-based decision-making is important, especially for market or policy transformation (Ng 2010; Mills et al. 2010). This understanding is only possible with scientific and parametric studies that examine the sensitivity and performance of various design and planning options. The parameters should be defined carefully and practically to produce realistic results.

References

Ashie Y, Hirano K, Kono T (2009) Effects of sea breeze on thermal environment as a measure against Tokyo's urban heat island. Paper presented at the seventh international conference on urban climate, Yokohama, Japan

Betanzo M (2007) Pros and cons of high density urban environments. Build 39–40

Blocken B, Carmeliet J, Stathopoulos T (2007) CFD evaluation of wind speed conditions in passages between parallel buildings—effect of wall-function roughness modifications for the atmospheric boundary layer flow. Wind Eng Indus Aerodyn 95:941–962

Cheng V, Ng E (2006) Thermal Comfort in urban open spaces for Hong Kong. Architectural Sci Rev 49(2):179–182

Cheng V, Ng E, Chan C, Givoni B (2011) Outdoor thermal comfort study in a subtropical climate: a longitudinal study based in Hong Kong. Int J Biometeorol 13(4):586–594

Ferziger JH, Peric M (2002) Computational method for fluid dynamics, 3rd edn. Springer, Berlin

Fluent I (2006) FLUENT 6.3 User's Guide, pp 12–58

Hong Kong Planning Department (HKPD) (2005) Feasibility study for establishment of air ventilation assessment system, Final report. The government of the Hong Kong Special Administrative Region

Hong Kong Planning Department (HKPD) (2011) Hong Kong planning standards and guidelines. Hong Kong Special Administrative Region

Hoppe P (1999) The physiological equivalent temperature—a universal index for the biometeorological assessment of the thermal environment. Int J Biometeorol 43:71–75

Kleissl J, Parlange MB (2004) Field experimental study of dynamic smagorinsky models in the atmospheric surface layer. J Atmos Sci 61:2296–2307

Kondo H, Asahi K, Tomizuka T, Suzuki M (2006) Numerical analysis of diffusion around a suspended expressway by a multi-scale CFD model. Atomospheric Environ 42(38):8770–8784

Kuznik F, Brau J, Rusaouen G (2007) A RSM model for the prediction of heat and mass transfer in a ventilated room. Build Simul 919–926

Lawson T (2001) Building aerodynamics. Imperical College Press, London

Letzel MO, Krane M, Raasch S (2008) High resolution urban large-eddy simulation studies from street canyon to neighbourhood scale. Atmos Environ 42(38):8770–8784

Meneveau C, Katz J (2000) Scale-Invariance and turbulence models for large-eddy simulation. Annu Rev Fluid Mech 32:1–32

Menter FR, Kuntz M, Langtry R (2003) Ten years of industrial experience with the SST turbulence model. Turbul Heat Mass Transfer 4:625–632

Mills G, Cleugh H, Emmanuel R, Endlicher W, Erell E, McGranahan G, Ng E, Nickson A, Rosenthal J, Steemers K (2010) Climate information for improved planning and management of mega cities (needs perspective). Procedia Environ Sci 1:228–246

Mochida A, Murakami S, Ojima T, Kim S, Ooka R, Sugiyama H (1997) CFD analysis of mesoscale climate in the Greater Tokyo area. J Wind Eng Ind Aerodyn 67–68:459–477

Mochida A, Tominaga Y, Murakam S, Yoshie R, Ishihara T, Ooka R (2002) Comparison of various κ-ε models and DSM applied to flow around a high-rise building—report on AIJ cooperative project for CFD prediction of wind environment. Wind Struct 5(2–4):227–244

Moss P (2008) Hong Kong an affair to remember. FromAsia Books Limited, Hong Kong

Murakami S (2006) Environmental design of outdoor climate based on CFD. Fluid Dyn Res 38(2–3):108–126

Murakami S, Deguchi K (1981) New criteria for wind effects on pedestrians. Wind Eng Indus Aerodyn 7:289–309

Ng E (2009a) Designing high-density cities. Earthscan, London Sterling

Ng E (2009b) Policies and technical guidelines for urban planning of high density cities—air ventilation assessment (AVA) of Hong Kong. Build Environ 44 (1478–1488)

Ng E (2010) Towards a planning and practical understanding for the need of meteorological and climatic information for the design of high density cities—a case based study of Hong Kong. Int J Climatol

Ng E, Cheng V, Chan C (2008) Urban climatic map and standards for wind environment—feasibility study. Technical input report no. 1: Methodologies and finds of user's wind comfort level survey. Hong Kong Planning Department, Hong Kong

Ng E, Yuan C, Chen L, Ren C, Fung JCH (2011) Improving the wind environment in high-density cities by understanding urban morphology and surface roughness: a study in Hong Kong. Landscape Urban plan 101(1):59–74

Oke TR (2006) Initial guidance to obtain representative meteorological observations at urban sites. World Meteorological organization, Switzerland

Stathopoulos T, Storms R (1986) Wind environmental conditions in passages between buildings. Wind Eng Indus Aerodyn 95:941–962

To AP, Lam KM (1995) Evaluation of pedestrian-level wind environment around a row of tall buildings using a quartile-level wind speed descriptor. Wind Eng Indus Aerodyn 54–55:527–541

Tominaga Y, Mochida A, Yoshie HK, Nozu T, Yoshikawa M, Shirasawa T (2008) AIJ guidelines for practical applications of CFD to pedestrian wind environment around buildings. J Wind Eng Ind Aerodyn 96:1749–1761

United Nations Population Division (2006) World urbanization prospects: the 2005 revision. United Nations, New York
Waller J (2008) Hong Kong the growth of the city. Compendium Publishing, London, United Kingdom
Wikipedia (2011) Mong Kok. http://en.wikipedia.org/wiki/Mong_Kok
Wolf M (2009) Hong Kong outside. Asia One Books and Peperoni Books, Hong Kong
Wong FJ (2004) Computational fluid dynamics analysis. Tsinghua University Press, Beijing
World Wildlife Fund (WWF) (2010) Hong Kong ecological footprint report 2010—paths to a sustainable future
Yim SHL, Fung JCH, Lau AKH (2009a) Mesoscale simulation of year-to-year variation of wind power potential over southern China. Energies 2:340–361
Yim SHL, Fung JCH, Lau AKH, Kot SC (2009b) Air ventilation impacts of the "wall effect" resulting from the alignment of high-rise buildings. Atomospheric Environ 43(32):4982–4994

Chapter 6
Natural Ventilation Modeling and Analysis for Climate-Sensitive Architecture Design

6.1 Introduction

Due to complicated building geometries and heterogeneous urban morphologies, this chapter introduces a project-specific wind analysis using Computational Fluid Dynamics (CFD) simulation to identify outdoor natural ventilation issues for any particular projects and develop appropriate mitigation strategies to address these issues. Knowledge-based decisions in the design process are significantly important for the mitigation of negative impacts of proposed buildings on the surrounding natural ventilation performance. Scientific steps taken to support knowledge-based decisions in the design process are (1) simplify local wind availability to identify the boundary condition for wind environment modeling at the neighborhood and building scales; (2) model airflow adjacent to the proposed buildings; (3) collect wind data that captures both local and global the wind environment; and (4) analyze modeling results. While the methods for the above four steps might be suitable for research works, they are less effective for the design practices due to more complicated inputs boundary condition, urban context, building typologies, divergent requirements and interests between architects, wind engineers, and researchers. As a result, Hong Kong Planning Department (HKPD) stated in Technical Circular No. 1/06 for improving natural ventilation performance (HKPD 2005) that "*CFD may be used with caution… There is no internationally recognized guideline or standard for using CFD in outdoor urban scale studies.*"

This chapter aims to present a practical application of the CFD simulation and addresses the issues raised in the Technical Circular No. 1/06 that are often encountered by wind consultants and architects. Figure 6.1 shows an exemplary procedural

Originally published in Chao Yuan and Edward Ng, 2014. Practical application of CFD on environmentally sensitive architectural design at high density cities: A case study in Hong Kong. Urban Climate, 8, pp. 57–77, © Elsevier, https://doi.org/10.1016/j.uclim.2013.12.001.

C. Yuan, *Urban Wind Environment*, SpringerBriefs in Architectural Design and Technology, https://doi.org/10.1007/978-981-10-5451-8_6

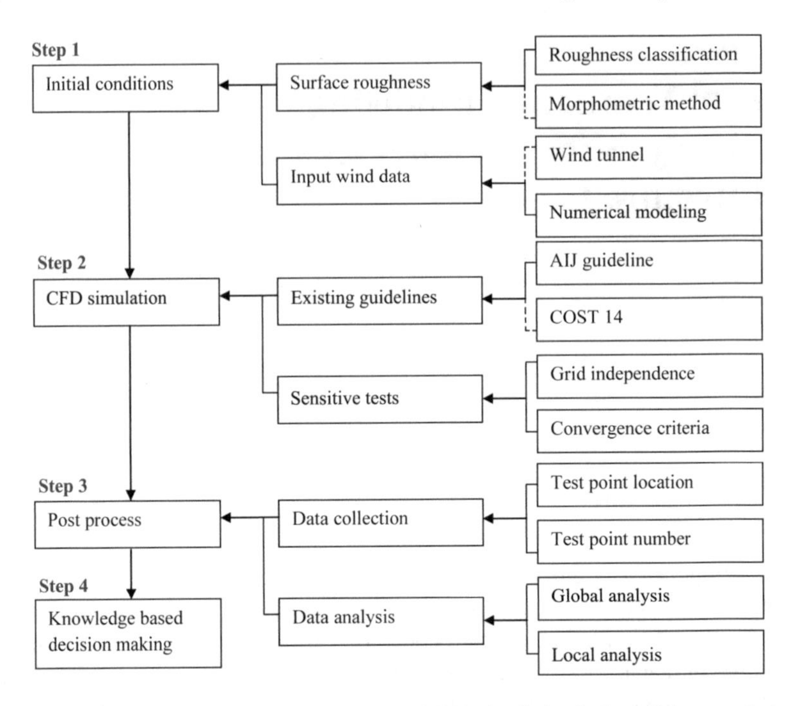

Fig. 6.1 Framework for the practical application of CFD simulation in the architecture design

framework for CFD simulation in an actual architectural design process that allows a more practical and dependable application, and streamlines the results to make them more meaningful for architectural design.

6.2 A Hong Kong Case Study

In this case study, the proposed building is rectangle in shape with small gap with the adjacent existing buildings, as shown in Fig. 6.2. Therefore, design strategies are needed to mitigate any potential negative effects of the proposed building on surrounding wind environment. Three design options with different mitigation strategies are provided by the Architecture Service Department (ASD). CFD simulation using ANSYS 13 (Adopted for meshing: ICEM; Adopted for computational fluid dynamics simulation: Fluent) is conducted to compare the mitigation strategies in these three options and identifies the optimal design option.

Three design options are presented in Fig. 6.3. In design option 1, two air passageways at the ground floor are provided and arranged at the north and south sides of the proposed building. Except for the low-zone passageways, building porosity in the upper floor is also applied in this design option. In design option 2, wider air passageways are provided at the ground floor. More importantly, an air passage is

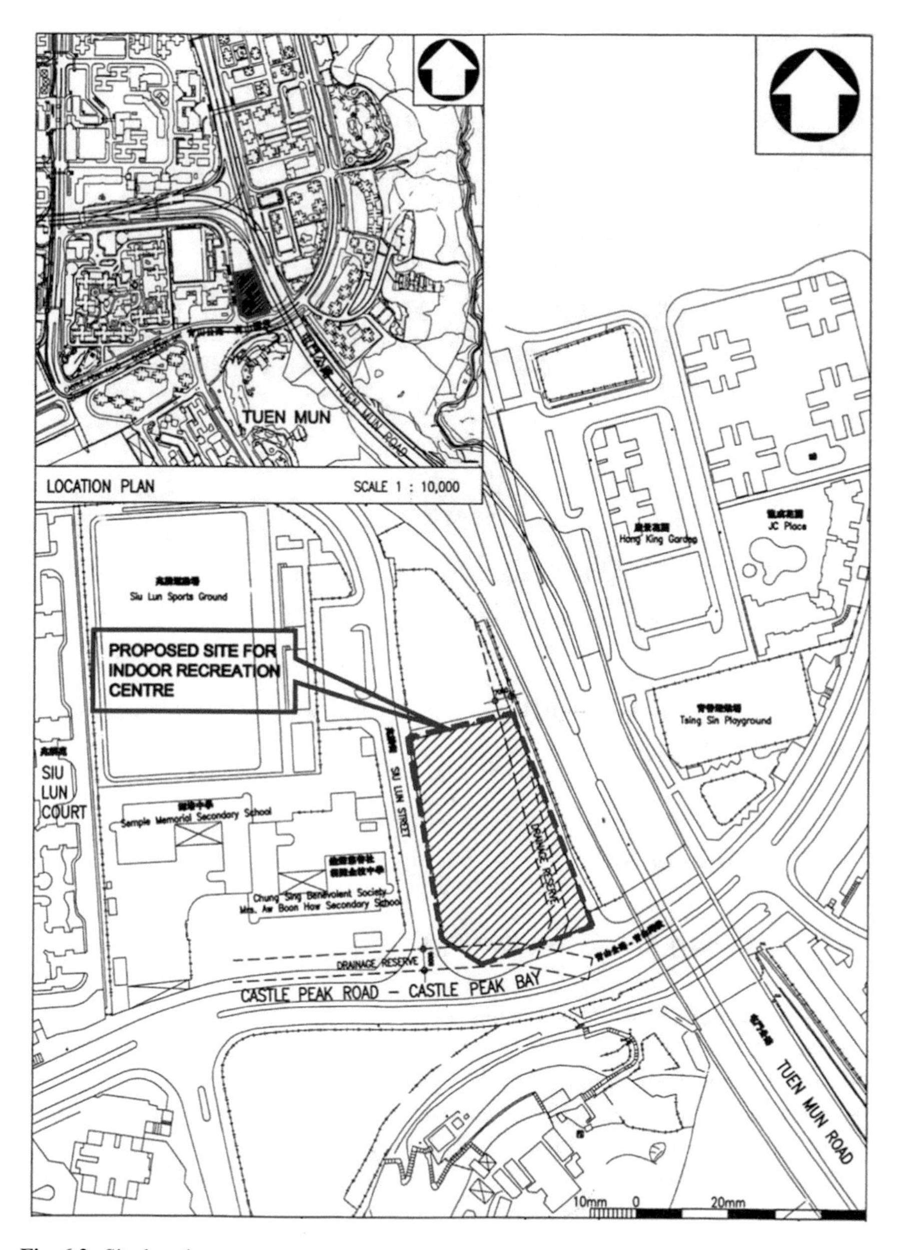

Fig. 6.2 Site location

arranged at the middle of the proposed building, where building permeability could be more efficient to improve the natural ventilation at leeward areas. The design option 3 is established based on design option 2. Apart from the different details on

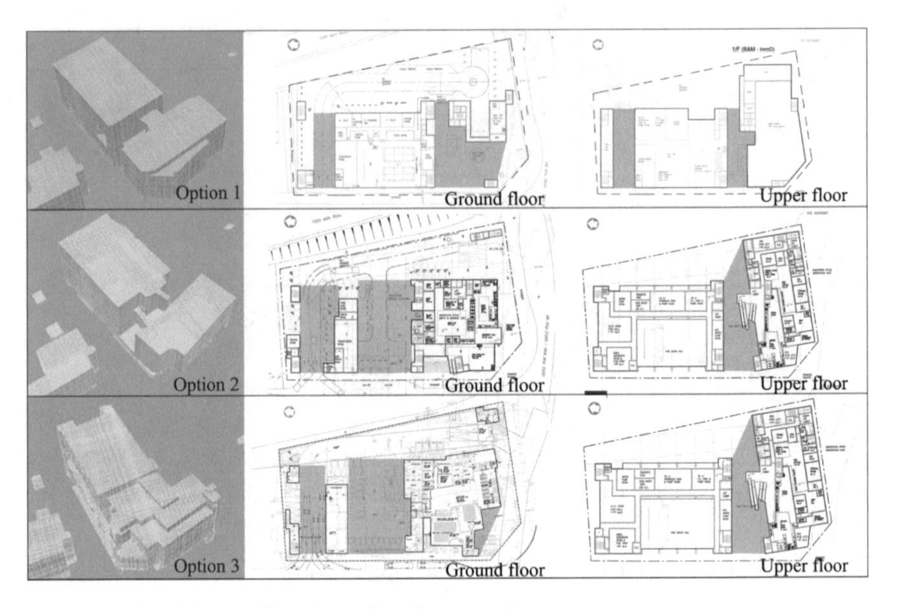

Fig. 6.3 Design options with various mitigation strategies

building morphologies in this design option, relatively wider air passageways and larger building porosity are included.

6.3 CFD Simulation

6.3.1 Boundary Condition Settings

Annual wind data was derived from a coupled Fifth-Generation NCAR/PSU Mesoscale Model (MM5)/CALMET model (Yim et al. 2007), and it was simplified to be used as input wind data in the CFD simulation. Annually averaged wind speed at the reference height (120 m) and wind probability in seven prevailing wind directions are highlighted in Table 6.1. These major prevailing wind directions are selected based on the requirement from the Technical Circular No. 1/06: "*the probability of wind coming from the reduced set of directions should exceed 75% of the time in a typical reference year*" (Hong Kong Building Department (HKBD) 2006). Surface roughness is set based on the classification in Architectural Institute of Japan (AIJ) guidelines (Tominaga et al. 2008). The input vertical wind speed profile can be set as

$$V_{(h)} = V_{\text{met}} \cdot \left(\frac{h}{d_{\text{met}}}\right)^{\alpha}, \tag{6.1}$$

Table 6.1 Annual wind availability data on site from MM5 model (The selected prevailing wind directions are highlighted)

Wind directions (i)	Mean wind speed (m/s)	Wind probability (%)
N (0°)	7.1	3.90
NNE (22.5°, $i = 1$)	**10.6**	**13.10**
NE (45°, $i = 2$)	**9.5**	**14.30**
ENE (67.5°, $i = 3$)	**6.1**	**7.80**
E (90°, $i = 4$)	**5.7**	**5.20**
ESE (112.5°, $i = 5$)	**7.2**	**11.80**
SE (135°, $i = 6$)	**6.2**	**13.00**
SSE (157.5°, $i = 7$)	**6.0**	**9.90**
S (180°)	5.2	5.10
SSW (202.5°)	4.8	4.40
SW (225°)	5.7	3.70
WSW (247.5°)	5.0	2.20
W (270°)	4.8	1.10
WNW (292.5°)	5.2	0.80
NW (315.5°)	6.2	1.50
NNW (337.5°)	6.3	1.50

Total selected wind probability larger than 75%

where α is the surface roughness factor, 0.35, V_{met} is the mean wind speed from an individual wind direction at 120 m (Table 6.1), and d_{met} is the reference height, 120 m above the ground. The outflow boundary condition is set as a pressure boundary condition with dynamic pressure equal to zero.

In addition, surrounding buildings and terrain (Fig. 6.4) are included in a 400 × 400 m modeling area to downscale the input wind information from urban scale (annual wind data from MM5) to the neighborhood scale. Thus, the input wind profiles (i.e., wind speed and turbulence intensity profiles) can be further modified, and the impact of target building on surrounding wind environment can be accurately modeled.

6.3.2 *Modeling Settings—Domain Size*

The computational domain for all design options is presented in Fig. 6.5. Blockage ratio (R_{b}) is calculated as

$$R_{\text{b}} = V_{\text{model}} / V_{\text{domain}}, \tag{6.2}$$

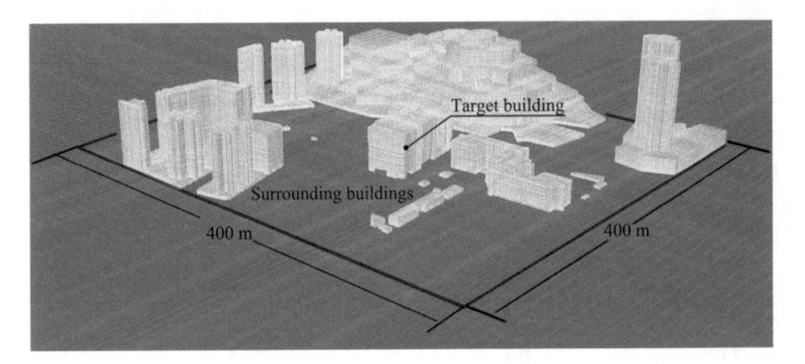

Fig. 6.4 Close-up of the modeling area

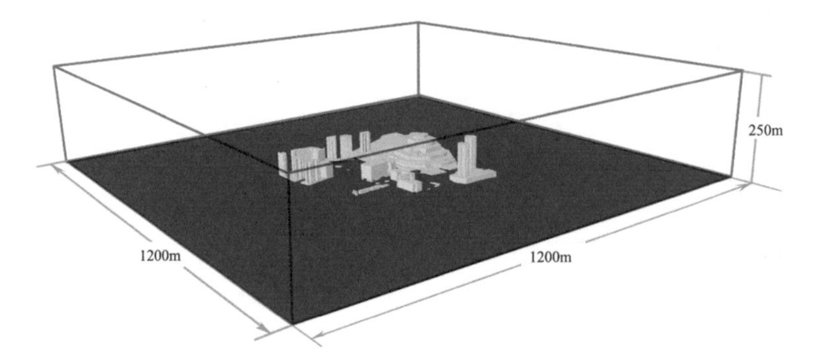

Fig. 6.5 Computational domain size

where V_{domain} and V_{model} are the domain and model volumes, respectively. V_{model} is estimated by the area covered by the terrain and buildings as 91.204 m^2, and thus the blockage ratio is 2.5%, assuming the average building height is 100 m. Considering that the height of most buildings is less than 100 m, the blockage ratio in these cases is less than 3%, which is considered small enough to avoid the impact of domain boundaries on airflow.

6.3.3 Modeling Settings—Grid Resolution

To efficiently use the computational capability, fine grids are generally arranged at a 400 × 400 m modeling area, whereas coarse grids are arranged at less important surrounding areas. Based on AIJ guidelines, three layers with height of 0.5 m each are arranged below the evaluation height (fourth layer, 2 m above the ground), as shown in Fig. 6.6. The maximum grid size ratio is set to 1.3, as required by the AIJ guidelines.

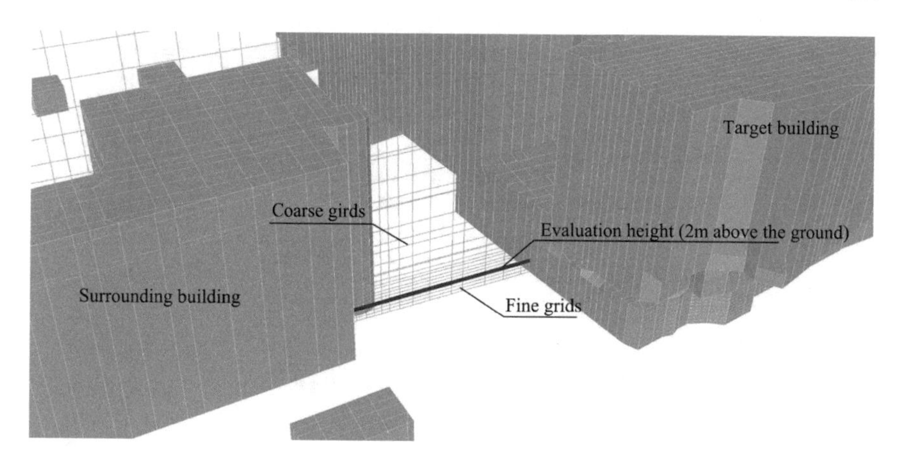

Fig. 6.6 First four layers parallel to the ground surface and evaluation height

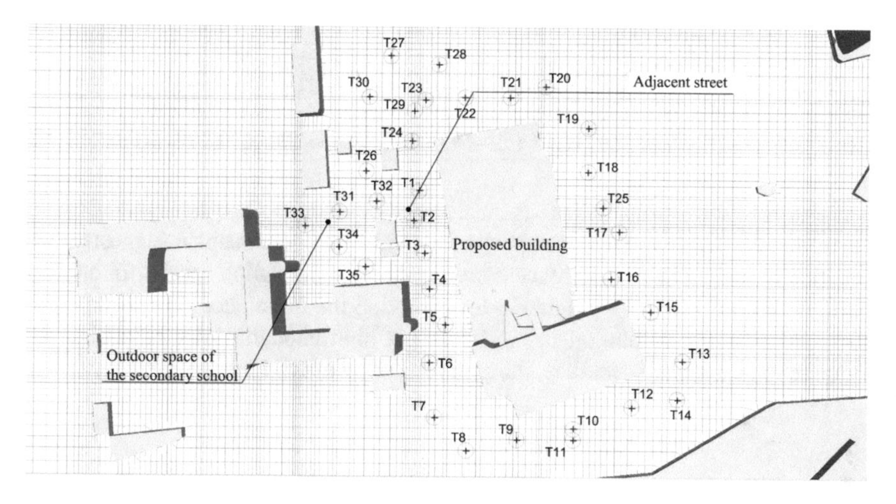

Fig. 6.7 Test points (sample size $n = 35$)

As the framework suggested by Frank et al. (2012), the successively refined resolutions are tested for grid dependency. Three simulations that only differ by grid point number (Grid 1: 7.0 million, Grid 2: 5.5 million, and Grid 3: 3.0 million) are conducted for grid dependence tests. In this matched-pair group, wind speeds are collected at 35 test points plotted at Fig. 6.7. The number of test points is determined to comply with the requirements for constructing robust hypothesis tests. More details on test point locations are introduced in the Result Analysis section.

Inferential statistics are conducted using t-distribution to test the sensitivity of simulation results on the change of grid resolution. As tabulated in Table 6.2, a *p*-value of 0.698 in grid resolution 1–2 indicates no difference between the wind data simulated in the resolutions of Grid 1 and 2 at the significance level of 0.05. However,

Table 6.2 Paired samples hypothesis test (significant level: 0.05; sample size: $n = 35$)

	Paired differences					t	Sig. (2-tailed) (p-value)
	Mean	Std. deviation	Std. error mean	95% confidence interval of the difference			
				Lower	Upper		
Grid resolution 1–2	0.00543	0.08219	0.01389	−0.02280	0.03366	0.39	0.698
Grid resolution 1–3	0.06508	0.20725	0.03503	−0.00612	0.13627	1.85	0.072

Table 6.3 Different scaled residuals in simulations

	Continuity	U-x	U-y	U-z	Energy	k	Omega
Criterion-1	4.26E−05	1.13E−06	1.08E−06	6.07E−07	5.10E−08	1.35E−05	1.97E−06
Criterion-2	2.70E−04	7.77E−06	7.56E−06	4.49E−06	5.64E−08	7.72E−04	1.71E−05
Criterion-3	1.01E−03	7.76E−05	8.16E−05	5.26E−05	5.82E−08	1.27E−03	9.04E−05

a p-value of 0.072 in the hypothesis test for Grid 1 and 3 indicates a suggestive but statistically insignificant difference between the two simulation results. In another word, both Grid 1 and 2 are suitable for resolving the turbulence flow in this case study, whereas the resolution in Grid 3 is not high enough. After balancing the computational cost and accuracy, the grid resolution in Grid 2 is adopted.

6.3.4 Modeling Settings—Convergence Criteria

Subsequently, RANS κ–ω SST (Shear Stress Transport) model, which comprises the standard κ–ω model (near-wall model) and the κ–ε model (the model for the outer part of the near-wall region) (Menter et al. 2003; Fluent Inc. 2012) is used. The performance of κ–ω SST model is evaluated by comparing the simulation results with existing wind tunnel data (2:1:1 shape building model) from AIJ in Chap. 5. Three simulations with different convergence criterions are conducted to test iterative convergence independence. In these three simulations, the scaled residuals for continuity, U-x, U-y, and U-z, as well as energy κ and ω, are given in Table 6.3. Inferential statistics similar with statistical analysis performed for grid independence are conducted.

Analysis results are tabulated in Table 6.4. A p-value of 0.324 in criteria 1–2 indicates no difference between two simulation results at the significance level of 0.05. These results indicate that the simulation results are independent from criteria

Table 6.4 Paired samples hypothesis test (significant level: 0.05, sample size $n = 35$)

	Paired differences					t	Sig. (2-tailed) (p-value)
	Mean	Std. deviation	Std. error mean	95% confidence interval of the difference			
				Lower	Upper		
Criteria 1-2	−0.00001	0.00005	0.00001	−0.00003	0.00001	−1.000	0.324
Criteria 1-3	−0.05371	0.11448	0.01935	−0.09304	−0.01439	−2.776	0.009

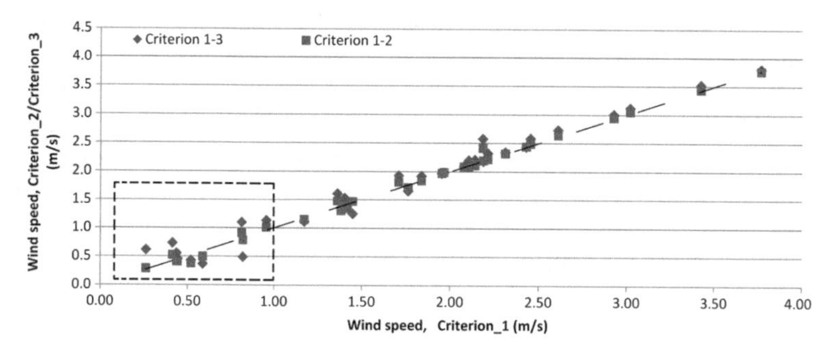

Fig. 6.8 Linear regression analysis to investigate the impact of convergence criteria on the accuracy of simulation

1 and 2. But a p-value of 0.009 in the hypothesis test for criteria 1–3 indicates a significant difference between two simulation results at the significance level of 0.05.

To further clarify the impact of convergence criteria on simulation accuracy, a regression analysis is conducted, and results are shown in Fig. 6.8. The larger variance highlighted in the low wind speed area (<1 m/s) indicates that the simulation with the convergence criterion 3 fails to reproduce low-speed airflow. Since the evaluation on low-speed airflow is critically important in this case study, the scaled residuals in all of the scenarios have to be equal to or smaller than the ones in criterion 2 in order to ensure result accuracy, especially for low-speed airflow.

6.4 Modeling Result Analysis

6.4.1 Global Analysis

The wind velocity contour maps of 21 scenarios (three design options × seven wind directions), partly shown in Fig. 6.9, indicate that all design strategies in the three design options are beneficial to the pedestrian-level wind environment surrounding the proposed building. Air can flow through the target building from the windward to the leeward areas. However, design option 1 is not as efficient as options 2 and 3, especially for promoting wind environment in the outdoor space of the secondary school and adjacent street (Fig. 6.9). It indicates that air passageways at the upper floor in design option 1 are not as efficient as that at the ground floor in design options 2 and 3 in terms of benefitting the pedestrian-level wind environment.

To further examine the above-mentioned modeling results, total 35 test points are selected at the pedestrian level for the statistical analysis, as shown in Fig. 6.10. Overall, the test points are evenly distributed and positioned in the surrounding open spaces, on the roads, and at places that pedestrians frequently access. The special test points are arranged in areas with strategic importance, such as test points 31–35 at the outdoor space of secondary school as shown in Fig. 6.7.

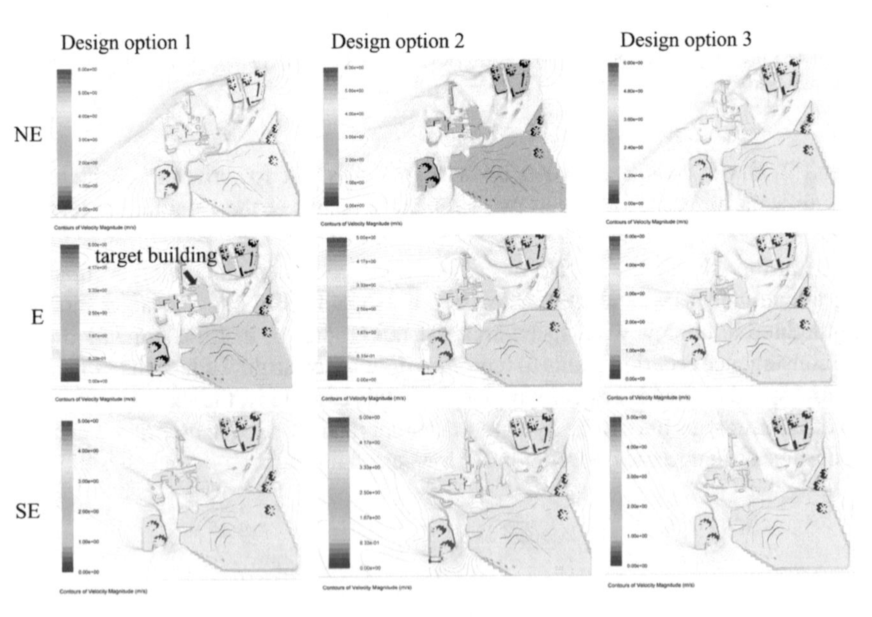

Fig. 6.9 Summary of simulation results, contours colored by wind speed in three design options in three input wind directions. (For interpretation of the references to color in the text, the reader is referred to the electronic version of this book)

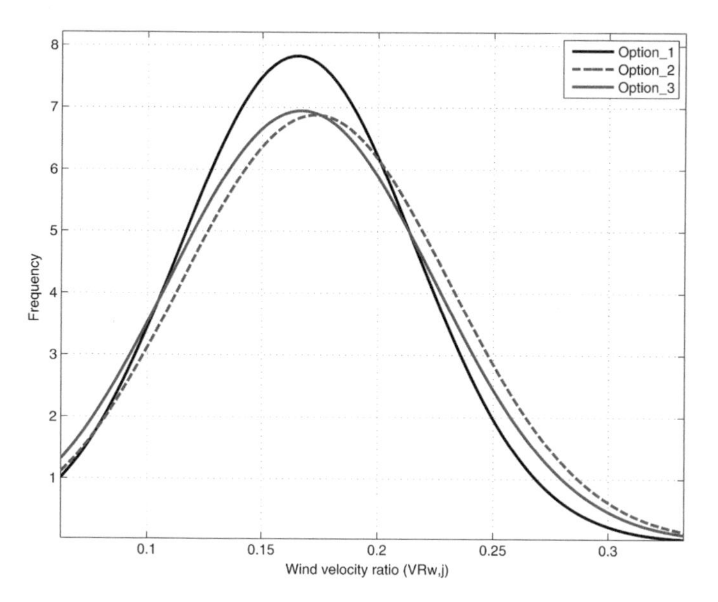

Fig. 6.10 Normal distribution fitting of $VR_{w,j}$ in design options 1, 2, and 3. (Significant level: 0.05)

Overall wind velocity ratio ($VR_{w,j}$) at the test points is calculated, and the normal distribution fittings of $VR_{w,j}$ for all three design options are presented in Fig. 6.10. The pedestrian-level wind environments in two design options 2 and 3 are similar and more efficient than that of design option 1.

6.4.2 Local Analysis

Rather than global analysis, we conduct a local analysis to further evaluate the wind environment within particular wind directions. Cross-comparison of $VR_{w,j}$, point by point, is conducted to clarify and compare the impact of design options 2 and 3 on natural ventilation performance.

The wind velocity ratio polar of the test points at the location with strategic importance (test points 1 and 31) is plotted in Tables 6.5 and 6.6. The wind polar is designed for seven inlet wind directions, and the lengths on the each direction represent the value of $VR_{w,j}$ in the different design options. Several important facts from the local analysis are as follows:

(a) As shown in Table 6.5, the wind polar at test point 1 represents the wind environment at the adjacent street in different wind directions. The wind velocity ratio in the input wind directions NE, NEE, E, and EES in design option 3 is significantly higher than those in design options 1 and 2. This result is not evi-

Table 6.5 Test points 1 and 31

Position	Wind velocity ratio polar (total 75.1% annual wind availability)

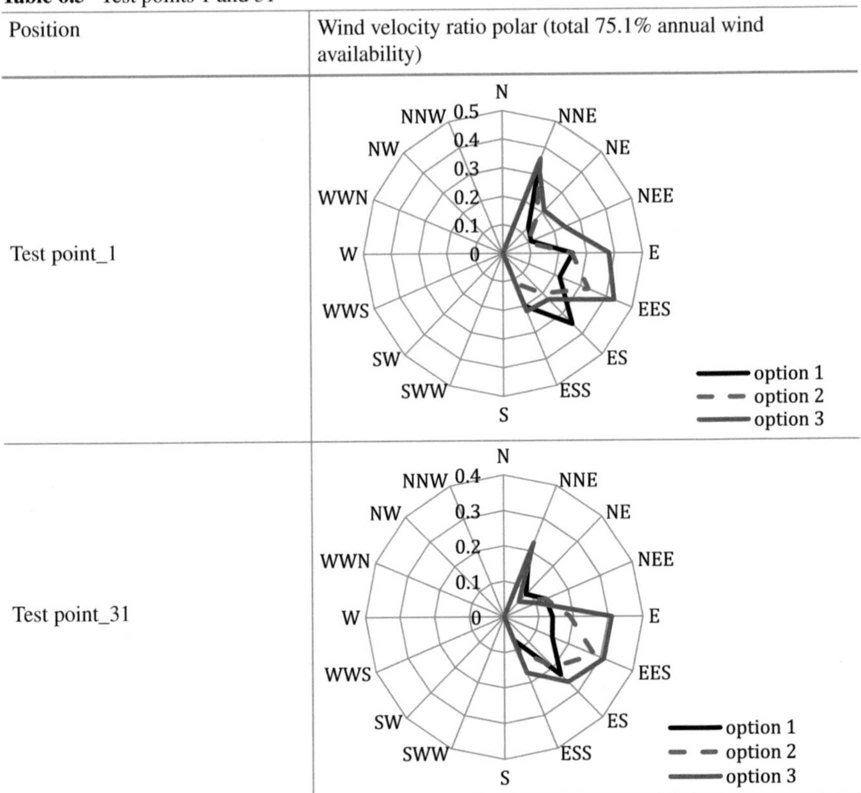

Special for seven inlet wind directions and the length on each direction means the respective value of VR

Table 6.6 Wind velocity rose at test point 24

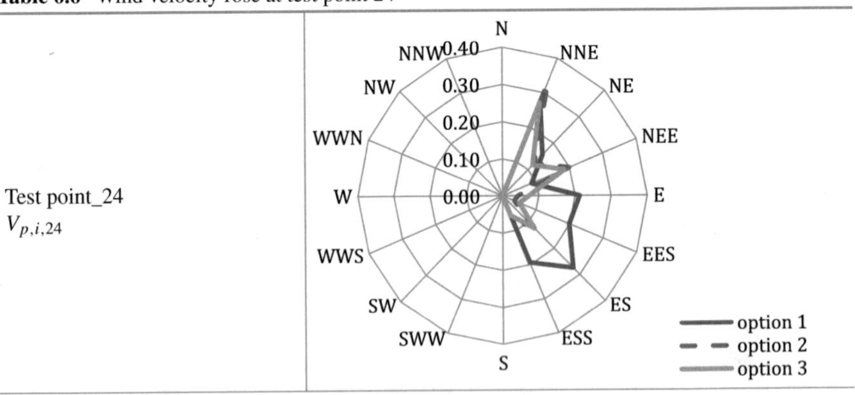

Special for seven inlet wind directions and the length on each direction means the respective value of $V_{p,i,j}$

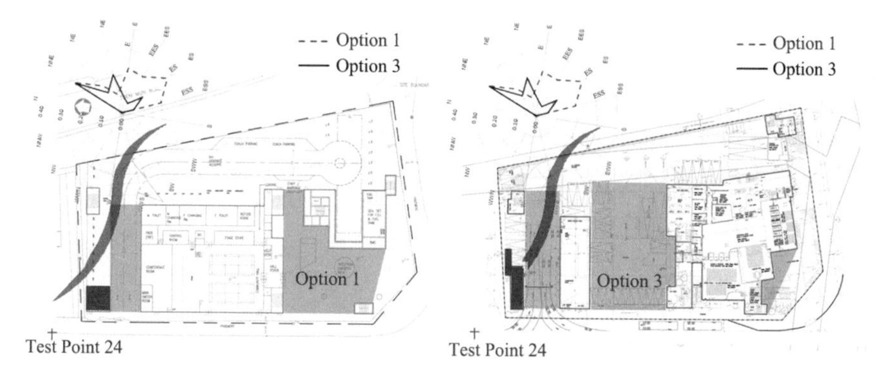

Fig. 6.11 Wind velocity ratios polar at the test point 24 and ground floor plan are presented together to illustrate how to identify the part of the proposed building in option 3 that could block the incoming air flow

dent in the averaged global analysis, in which the performances of options 2 and 3 are similar.

(b) The wind polar at test point 31 indicates the efficiency of the different design option strategies in terms of natural ventilation performance at the outdoor space in the secondary school. The wind velocity ratios in the input wind directions E, EES, ES, and ESS indicate that the air can flow through the proposed building the earliest upon adopting mitigation strategies in design option 3. Based on the above global and local analyses, the performances of options 2 and 3 are found to be globally better than that of option 1. At the location with strategic importance, option 3 outperforms option 2.

(c) After identifying the optimal design option, a more detailed understanding can be obtained to further improve the optimal design option. The wind velocity ratio polar in Table 6.6 indicates that the natural ventilation performance at test point 24 in design option 3 can be significantly poorer than that in design option 1. As shown in Fig. 6.11, combining the wind polar and ground floor planning, the limitation can be attributed to the part of the building in option 3 that is highlighted by in red. An appropriate modification based on this understanding can significantly improve the annual averaged natural ventilation performance of this area.

Based on the foregoing modeling and sufficient statistical analysis, the effects of different design options on the surrounding wind environment are well evaluated and cross-compared. Practical and accurate insights about optional design option identification and improvement are provided for architects. All mitigation strategies in design options 1, 2, and 3 are found to be helpful in mitigating the negative effects of the proposed building on the surrounding wind environment. However, the mitigation measurement in option 1 is not as sufficient as that in options 2 and 3. By adopting the mitigation strategies in option 2 or 3, natural ventilation performance at the surrounding areas, especially at the adjacent street and the secondary school, can be significantly improved. Furthermore, the performance of option 3 is found to be better than that of option 2 at several locations with strategic importance based on the wind polar analysis. Therefore, the design option 3 is considered as the optimal design option and is thus recommended. Detailed modification recommendations for the design option 3 are also provided based on wind velocity ratio polar analysis.

6.5 Conclusion

This chapter focuses on a practical application of CFD simulation to bridge the gap between wind engineering and architectural design using a Hong Kong case study. To ensure the accuracy of CFD simulation in a real case study, apart from the setting requirements mentioned in the AIJ guidelines, we provide statistical methodology for the grid dependence and convergence criteria sensitivity test to validate modeling accuracy in practical applications. The application of hypothesis testing and regression analysis makes the conclusions of the sensitivity tests more robust and confident. For the practical CFD application, this statistical method is much more practical than validation method (cross-comparing with wind tunnel data) applied in the existing research works. The analysis results indicate that a reduction in the scaled residuals of at least three orders of magnitude could be acceptable in practical applications. Furthermore, this study emphasizes the importance of data collection and analysis in the implementation of architectural design. The new data analysis process is presented from the global analysis based on referential statistics to the local analysis based on the wind velocity ratio polar. Through this analysis process, reliable understandings of the relationship between the surrounding natural ventilation performances and building morphologies can be obtained by architects to easily identify critical design indexes and to make the corresponding design decisions.

References

Fluent Inc. (2012) FLUENT 14.0 theory guide

Frank J, Sturm M, Kalmbach C (2012) Validation of OpenFOAM 1.6. x with the German VDI guideline for obstacle resolving micro-scale models. COST Office, Brussels, Belgium

Hong Kong Planning Department (HKPD) (2005) Feasibility study for establishment of air ventilation assessment system, Final report. The government of the Hong Kong Special Administrative Region, Beijing

Hong Kong Building Department (HKBD) (2006) Sustainable building design guidelines. Practical note for authorized persons, registered structure engineers and registered geotechnical engineers. APP-152, Hong Kong

Menter FR, Kuntz M, Langtry R (2003) Ten years of industrial experience with the SST turbulence model. Turbul Heat Mass Transfer 4:625–632

Tominaga Y, Mochida A, Yoshie R, Kataoka H, Nozu T, Yoshikawa M, Shirasawa T (2008) AIJ guidelines for practical applications of CFD to pedestrian wind environment around buildings. J Wind Eng Ind Aerodyn 96:1749–1761

Yim SHL, Fung JCH, Lau AKH, Kot SC (2007) Developing a high-resolution wind map for a complex terrain with a coupled MM5/CALMET system. J Geophys Res 112:D05106

Part IV
Others—Urban Air Quality and Trees

Chapter 7
Improving Air Quality by Understanding the Relationship Between Air Pollutant Dispersion and Building Morphologies

7.1 Introduction

7.1.1 Background

People living in high-density cities suffer from both short- and long-term exposure to ambient air pollution, which has been associated with cardiorespiratory morbidity and mortality, mental health, and cancers (World Health Organization (WHO) 2008; European Environment Agency (EEA) 2012; Kunzli et al. 2000). Given to its ubiquitous exposure, the impact of air pollution on public health is of great significance (Kunzli et al. 2000). Vehicle emission (fuel combustion) is one of the main contributors of air pollution in the urban areas, particularly at street canyon levels. Major criteria air pollutants emitted directly from vehicle exhaust include particulate matter (PM), nitrogen dioxide (NO_2), carbon monoxide (CO); road traffic also emits other pollutants such as heavy metals, benzene (C_6H_6), and benzopyrene (BaP) (EEA 2012). To guide air pollution control effort, the WHO (2008) has explicitly recommended the concentration limits for each criteria air pollutant.

While an improved vehicle emission control program was implemented in Hong Kong by the Hong Kong SAR Government to decrease vehicle and road traffic pollution, the roadside concentration of NO_2 continues to increase in spite of the implantation of the program (Hong Kong Environmental Protection Department (HK EPD) 2011). Hourly, daily, and annual average concentrations of NO_2 at the roadside stations in the Central, Causeway Bay, and Mong Kok areas have far exceeded the

---

Originally published in Chao Yuan, Edward Ng and Leslie K. Norford, 2014. Improving air quality in high-density cities by understanding the relationship between air pollutant dispersion and urban morphologies. Building and Environment, 71, pp. 245–258, © Elsevier, https://doi.org/10.1016/j.buildenv.2013.10.008.

C. Yuan, *Urban Wind Environment*, SpringerBriefs in Architectural Design and Technology, https://doi.org/10.1007/978-981-10-5451-8_7

Fig. 7.1 Vehicle fleets in the deep street canyons of Mong Kok and Wan Chai in Hong Kong; high concentration of NO_2 is frequently measured at the roadside stations in the areas

WHO guidelines (HK EPD 2011). Those areas are high-density metropolitan areas and traffic hotspots. As shown in Fig. 7.1, vehicles crowd the streets of high-density urban areas in Hong Kong. Similar phenomenon has also been reported in Europe (EEA 2012). The higher concentration of roadside NO_2 reported may be the result of larger NO_2 percentage in total traffic emissions (EEA 2012; Grice et al. 2009) and poorer urban air ventilation in high-density urban areas (Ng et al. 2011; Yim et al. 2009b). Bulky building blocks, compacted urban volumes, and very limited open spaces significantly block the pollutant dispersion in the deep street canyons (Hang et al. 2012; Tominaga and Stathopoulos 2012). Therefore, apart from having control measures to decrease vehicle emissions, understanding pollutant dispersion as related to the urban planning and design mechanism is necessary in order to guide policymakers, planners, and architects in making better evidence-based decisions.

Urban microclimate and air quality can act as both the determinants and consequences of urban planning and design activities. Thus, research on the environmental sensitivity of urban planning and design is needed to delineate such interchangeable relationship. For example, Mirzaei and Haghighat (2012) provided a systematic approach to quantify the ambient environment in the street canyon. Huang et al. (2008) studied an actual urban case of mid-density layouts. Hang et al. (2012) examined the effects of varying building heights on street-level air pollutant dispersion, and Eefents et al. (2013) reported the effects of canyon indicators such as Sky View Factor (SVF) on the concentration of NO and NO_x. Moreover, Buccolieri et al. (2010) clarified the influence of building packing density on air pollution concentration. Richmond-Bryant and Reff (2012) linked the fluid properties and canyon geometries (e.g., the Reynolds number and canyon height) with air pollutant retention using field measurement data at Manhattan and Oklahoma. Despite the existing research, much remains unknown regarding the direct understandings on how to decrease air pollution concentration through urban morphological mechanisms, particularly for high-density cities, and the application of the understandings in practical design strategies. As a result, more parametric studies are needed to bridge the knowledge gap between high-density urban design and air pollutant dispersion mechanisms in the urban street canyons, thereby providing useful guidance to planners, designers, and policymakers.

7.1.2 Literature Review

Over the past decades, research on urban climate had focused on air quality in the street canyons (Li et al. 2006). For example, some studies conducted the "car-chasing" experiment and tunnel testing (Kristensson et al. 2004; Ho et al. 2009) to determine the real-world traffic emission factors of particles and gaseous pollutants, and to evaluate the chemical compositions of emissions from different vehicles. Not only did these studies provide important information on the public health impact of different air pollutants and inform policy guidelines on transportation control measures, findings from these studies also act as key references of input boundary conditions for pollutant dispersion modeling.

In addition to the "car-chasing" experiment and tunnel testing, computational fluid dynamics (CFD) studies of different scales was conducted to identify the pollutant dispersion phenomenon in idealized street canyons with a point or line emission source with (Baik et al. 2007; Kikumoto and Ooka 2012) or without chemical reaction (Salim et al. 2011; Tominaga and Stathopoulos 2011, 2012; Gousseau et al. 2010; Pontiggia et al. 2010). There were also researches on isothermal and non-isothermal street canyon in which the effect of convective and turbulent mass fluxes on dispersion was examined (Gousseau et al. 2011; Huang et al. 2008). Collectively, these CFD studies have provided critical insight into the pollutant dispersion patterns, as well as the relationship among wind velocities, turbulence intensities, and pollutant dispersion; they also enabled the performance evaluation of their dispersion and turbulence models by cross-comparing the CFD modeling results with wind tunnel experiment data.

Building upon the previous modeling and field measurement work, this chapter describes a parametric study that evaluated the applicability and effectiveness of the Air Ventilation Assessment (AVA) TC-1/06 guidelines and the Sustainable Building Design (SBD)'s APP-152 guidelines on promoting air pollutant dispersion. Specifically, the chapter includes a description of: (1) a validation study that identified the optimal modeling method for the parametric study by evaluating the performance of both the RANS and LES models in modeling species transport, (2) a series of parametric studies that helped understand the relationship among building geometry, urban permeability, and air pollutant dispersion, and evaluated the effects of future urban developments, based on the current planning trend, on air pollutant dispersion in the street canyons. Evaluation was provided on the effectiveness of various mitigation strategies on promoting pollutant dispersion and improving air quality at urban areas. Such evaluation is beneficial especially given the limited land resources in high-density cities and the numerous planning and design restrictions for development projects. The chapter ends with a discussion of the parametric study findings, which provides practical and accurate insights at the beginning stage of the design practice, avoids mistakes that may not be reversible at the late stages of the design process, and ultimately facilitates a paradigm shift from the traditional experience-based ways of designing and planning to a more scientific, evidence-based process of decision-making, which is necessary to meet the needs of designing high-density cities (Ng 2012).

7.2 CFD Simulation

7.2.1 Eulerian Method for Species Transport Modeling

Eulerian method is adopted to investigate the effectiveness of different design strategies on promoting air pollutant dispersion. The Eulerian and Lagrangian methods have been widely used to model air pollutant dispersion (Salim et al. 2011; Wang et al. 2012; Zhang and Chen 2006). Unlike the Lagrangian method, which considers the chemical species as a discrete phase, Eulerian method considers species as a continuous phase, and it is solved based on a control volume, which is similar in form to that for the fluid phase (Wang et al. 2012; Zhang and Chen 2006). Therefore, Eulerian method is more practical for calculating the gaseous pollutant concentration. Another reason for choosing Eulerian method was that it requires less computational capability than Lagrangian method, which needs to track more than several million particles in the neighborhood-scale computational domain for outdoor environment simulation. Eulerian method has been validated by comparing the simulated data with wind tunnel experiment results (Salim et al. 2011; Wang et al. 2012).

Here, ANSYS Fluent was used to solve the unsteady convection–diffusion equation to predict the mass fraction of species Y_i (Fluent Inc. 2012):

$$\frac{\partial}{\partial t}(\rho Y_i) + \nabla \cdot \left(\rho \vec{v} Y_i\right) = -\nabla \cdot \vec{J}_\iota + S_i + R_i, \tag{7.1}$$

where S_i is the rate of the user-defined source term, i is the number of species, ρ is the air density, $\vec{v}$ is the overall velocity vector, t is time, and $\vec{J}_\iota$ is the mass diffusion in turbulent flows. Inlet diffusion is enabled to include the diffusion flux of species at the emission inlet. Because the simulation excludes chemical reaction, the rate of the product from chemical reaction (R_i) is set to zero. $\vec{J}_\iota$ in the turbulence is estimated as (Fluent Inc. 2012):

$$\vec{J}_{\mathrm{t}} = -\left(\rho D_{\mathrm{i,m}} + \frac{\mu_{\mathrm{t}}}{\mathrm{sc}_{\mathrm{t}}}\right) \nabla Y_i - D_{\mathrm{T,i}} \frac{\nabla T}{T}, \tag{7.2}$$

where $D_{\mathrm{i,m}}$ and $D_{\mathrm{T,i}}$ are the mass diffusion coefficient for species i and the thermal diffusion coefficient, respectively, sc_{t} is the turbulent Schmidt number (the default value 0.7 is used in this study), and μ_{t} is the turbulent viscosity.

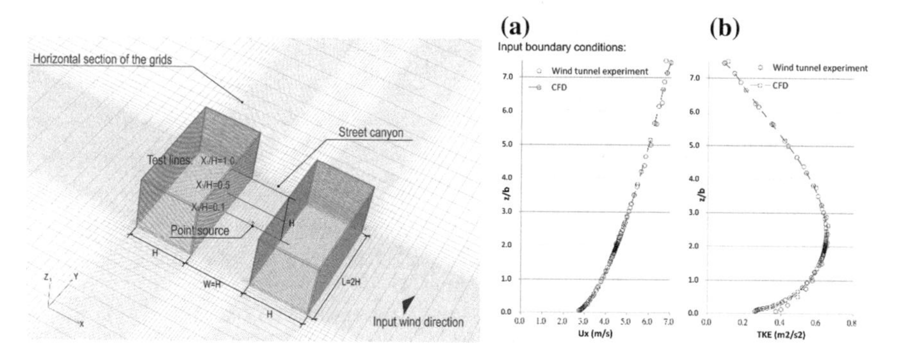

Fig. 7.2 Model configurations, Input conditions (U_x and TKE), and horizontal test lines in the street canyon; X_1, X_2, and X_3 are the heights of the test lines

7.2.2 Optimal Turbulence Model (Validation)

The wind tunnel data provided by Niigata Institute of Technology (Tominaga and Stathopoulos 2011) was used to validate the abovementioned air pollutant dispersion model. The performance of different turbulence models, including standard and realizable κ–ε model, Reynolds stress model (RSM), shear stress transport (SST) κ–ω model, and LES model, on air pollutant dispersion was also investigated in this validation study. Model configurations were set to match those in the wind tunnel experiment, as shown in Fig. 7.2 (Left). H/W and H/L aspect ratios were set to 1.0 and 0.5, respectively, where H is building height, W is the width of the street, and L is the length of the canopy. All modeling settings, such as domain size ($X \times Y \times Z$: $600 \times 200 \times 100$ m), grid type (structural grid), grid resolution (grid number: 3 million, and the grid size at street canyon was set at less than 1/10 of building length), grid size ratio (maximum grid size ratio was set at 1.3), followed the Architectural Institute of Japan (AIJ) guideline (Tominaga et al. 2008b). More than three grid layers were used under the test line, and the convergence criteria for turbulence and species were E-05 in the RANS model and E-04 in the LES modeling. The input wind direction was perpendicular to the street canyon. Input wind velocity (U_x) and turbulence kinetic energy (TKE) profiles plotted in Fig. 7.2 (Right) were set by a user-defined function. The surface roughness factor (α) was set at 0.21. The modeling settings are summarized at Table 7.1.

While the calculation in the RANS study is statistically steady, in the LES, study is often statistically unsteady. Thus, the time step in LES was set constant, i.e., $\Delta t = 0.001$ s, and the solution variables were time averaged until the flow became statistically steady, which was identified by monitoring the instantaneous values of variables at several points in the street canyon. The sampling time (t) was approximately 180 s. The Smagorinsky-Lilly model provided by ANSYS Fluent 14.0, where Cs = 0.12 as recommended by Tominaga et al. (2008a), was used in the LES modeling.

As shown in Fig. 7.2, the point source was arranged at the middle of the street canyon. Ethylene (C_4H_4), serving as a tracer gas, was released from the point source

Table 7.1 CFD simulation settings

Computational domain	X_Y_Z: 600_200_100 m
Blockage ratio	<5%
Grid expansion ratio	Less than 1.3
Grid resolution	Grid size less than 1/10 of building length (Fig. 7.2)
Grid number	3 million
Grid type	Structural grid (Fig. 7.2)
Prismatic layer	More than three grid layers under the test line
Inflow boundary condition	Input wind velocity (U_x) and turbulence kinetic energy (TKE) profiles reproduced by a user-defined function, based on the measurement at the wind tunnel experiment (Fig. 7.2)
Outflow boundary	Zero gradient condition
Near-wall treatment	Enhanced wall functions in κ-ε models and Reynolds stress model No wall functions in SST κ-ε and LES model
Solving algorithms	QUICK for momentum, TKE, and other convection terms
Relaxation factors	Default values in ANSYS Fluent 14.0
Convergence criteria	Below E_05 in RANS model Below E_04 in LES mode

with a wind velocity W_s ($W_s/U_b = 0.12$, U_b was the input wind velocity at the building height, 3.8 m/s). The tracer concentration was set at 1000 ppm, replicating the setting in the wind tunnel experiment.

Simulation results of different turbulent models were collected at three test lines shown in Fig. 7.2, and cross-compared in Fig. 7.3. While the computational cost of a LES model is several times higher than that of a RANS model, Tominaga and Stathopoulos (2011) reported that the emission concentration could be more accurately modeled by a LES model than by a RANS model, particularly at the windward area in the street canyon, because of the reproduction of the instantaneous fluctuation of the concentration.

Nonetheless, the performance of the RANS models provided by ANSYS Fluent 14.0 was significantly better than the ones reported by Tominaga and Stathopoulos (2011). The simulation results were closer to the experiment data, particularly at the bottom line ($X/H = 0.1$, X is the height of the test line). However, all RANS models overestimated the concentrations at the middle and top lines ($X/H = 0.5$ and 1, respectively), particularly at the windward side of the street canyon. Among the models, the SST κ–ω model (the red dash line) performed best in terms of air pollutant modeling at the windward side. The special near-wall region (i.e., the shear layer) treatment by the standard κ–ω model (Menter et al. 2003; Fluent Inc. 2012) was considered helpful for estimating the air pollutant concentration near surface regions. Overall, the SST κ–ω model offered acceptable accuracy and computational cost, and was considered to be a good tool for modeling air pollutant dispersion in the design process. Thus, the SST κ–ω model was selected as the preferred turbulence model to simulate the air pollutant dispersion of the cases in the parametric study.

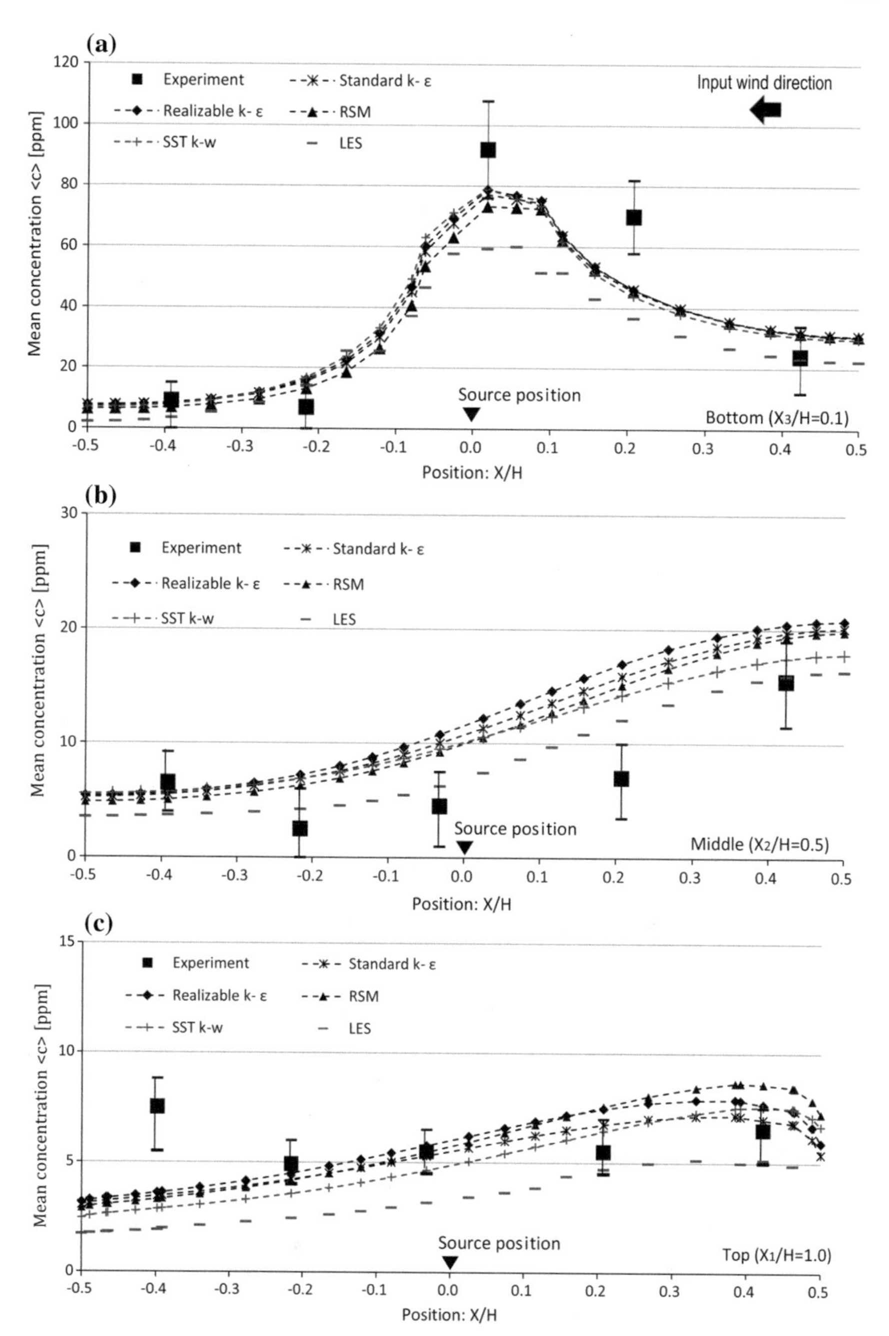

Fig. 7.3 Cross-comparisons of time-averaged concentrations <c> at the street canyon between the wind tunnel experiment and different turbulent models: **a** <c> at the bottom line ($X_3/H = 0.1$); **b** <c> at the middle line ($X_2/H = 0.5$); **c** <c> at the top line ($X_1/H = 1.0$). Error bars are the standard deviations of the measurements in the wind tunnel experiments

7.3 Parametric Study

Compared with other methods used in studies that focused on actual urban morphologies (Mochida et al. 1997; Kondo et al. 2006; Ashie et al. 2009), obtaining generic information for deriving guidelines is easier using parametric studies. By cross-comparing the sensitivity of chemical species dispersion to the parameter changes in various testing scenarios, critical design elements can be efficiently identified. For example, the results of the parametric study of the wing wall for room ventilation by Givoni (1969) were implemented by the Hong Kong Government (Ng et al. 2006). In order for the results of parametric studies to be relevant and applicable, parametric cases should resemble the reality (Ng et al. 2006). Therefore, a neighborhood-scale model was used in this study to obtain realistic parametric simulation results. As shown in Fig. 7.4, parametric geometric models were established based on the actual urban condition in Mong Kok, a high-density downtown area in Hong Kong, using a regular street grid. This geometric model realistically resembles airflows and air pollutant dispersion status in street canyons. Unlike previous parametric studies, where only one or two generic buildings or building arrays were included (Blocken et al. 2007; Gousseau et al. 2010; Salim et al. 2011; Stathopoulos and Storms 1986; Tominaga and Stathopoulos 2011, 2012; Yim et al. 2009b), effects of the urban context were included in this study.

As shown in Fig. 7.5, eight building geometries were designed to create corresponding parametric models with various building permeability. The details of parametric models were tabulated in Table 7.2. Cases 1 and 2 represented the current and future urban conditions, respectively, whereas Cases 3–8 were established based on the sustainable building design (SBD-APP-152) guidelines (Hong Kong Building Department (HKBD) 2006). In Cases 3–5, three design strategies—building setback (Strategy A), building separation (Strategy B), and stepped podium void (Strategy C)—were implemented, respectively. Building porosities were included in Case 6 (Special Strategy). For Case 7, building setback was combined with building separation. For Case 8, stepped podium void was combined with building separation. Plot ratios of Cases 3–8 were set to be similar to that of Case 2. For example, the decrease of floor area caused by incorporating mitigation strategies was compensated

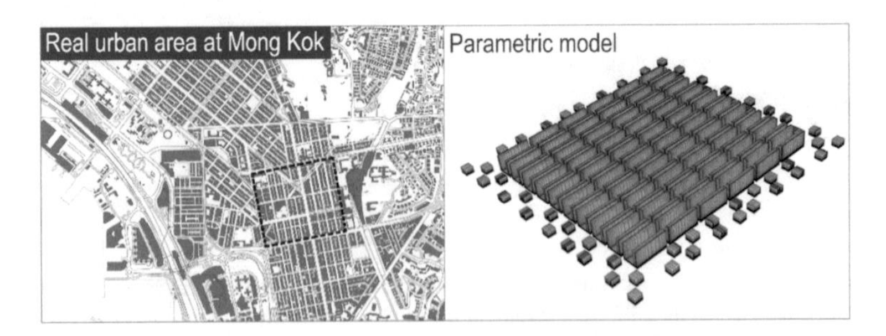

Fig. 7.4 Actual urban area located at Mong Kok and its corresponding parametric model

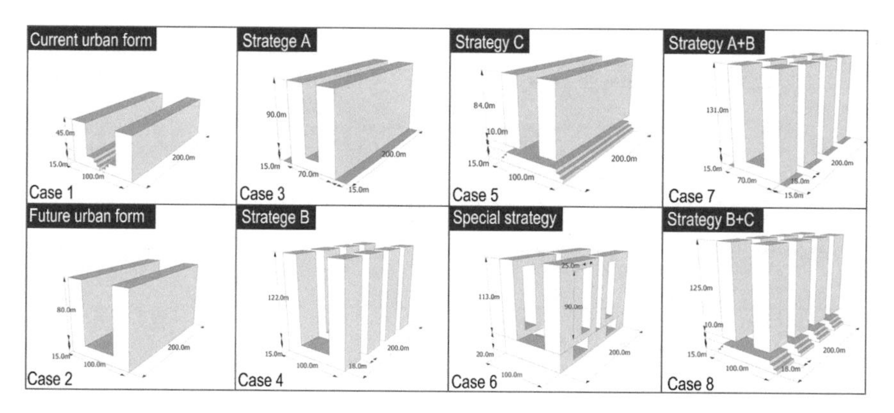

Fig. 7.5 Eight testing scenarios

by increasing the building height in Cases 3–8. The permeability of buildings (*P*) (HKBD 2006) and site area ratio (λ_p), which represent the vertical and horizontal permeability respectively, were calculated, and tabulated in Table 7.2. High values of *P* and λ_p indicate low permeability.

7.3.1 Modeling Settings in the Parametric Study

Eight parametric models were simulated using both the Eulerian species transportation model and SST κ–ω turbulence model to evaluate dispersion of chemical species in street canyons. As shown in Fig. 7.6, the computational domain size was 3.9 km × 4 km × 0.55 km ($X \times Y \times Z$). Other simulation settings were the same as the validation study.

To set the input wind velocity profile particularly for the Mong Kok area by means of the log law, the site-specific annual wind data (U_{met} = 11 m/s) at a 450 m height (d_{met}) were obtained from the fifth-generation NCAR/PSU mesoscale model (MM5) (Yim et al. 2009a). Given the high urban density of Hong Kong, the surface roughness factor (α) was set at 0.35.

NO_2 was selected as the emission gas of interest, because its concentration continues to increase in the metropolitan areas in Hong Kong. A line pollutant source ($X \times Y$: 1300 × 10 m) was set at the bottom of the street canyon in the middle of a building gap to represent the major road with heavy traffic volume at Mong Kok, as shown in Fig. 7.6. A reference emission NO_2 concentration ($\langle c_0 \rangle$ = 1000 ppm) was used. The input wind velocity ratio at the emission source followed the value used in the wind tunnel experiment (Tominaga and Stathopoulos 2011), W_s/U_b = 0.12. This study did not factor in the chemical reactions between NO_x and ozone (O_3) (Vardoulakis et al. 2003), as it aimed to determine how NO_2 could be dispersed and

Table 7.2 Eight testing scenarios in which air flows and species transport are simulated

Testing scenario	Case	Building geometry			Land use efficiency	Urban permeability		
		Parameter	Strategy style	H (m)	P	λ_p	λ_s	λ_i
1	Case 1	Current urban form	/	60	8.9	0.9	0.7	0.9
2	Case 2	Future urban form	/	95	14	0.9	0.8	0.9
3	Case 3	Building setback	Single	105	14	0.9	0.5	0.7
4	Case 4	Building separation	Single	137	14	0.7	0.6	0.5
5	Case 5	Stepped podium void	Single	109	14	0.8	0.6	0.7
6	Case 6	Building porosity	Single	133	14	0.7	0.8	0.7
7	Case 7	Building separation + setback	Multiple	146	14	0.7	0.4	0.3
8	Case 8	Building separation + stepped podium void	Multiple	150	14	0.6	0.4	0.4

Note P: Plot ratio; λ_p: permeability of buildings (vertical urban permeability); λ_s: site coverage ratio (horizontal urban permeability); λ_i: integrated permeability

transported by airflow in street canyons with different geometric parameters. The study was therefore restricted to the city spatial scale.

7.4 Result Analysis

This parametric study applied both contours of the normalized concentration and statistical analysis to evaluate the performance of design strategies on improving air quality at high-density urban areas, and to provide insight into the sustainable building design guidelines (HKBD 2006). This study used the normalized concentration ($\overline{c}$) as the index to analyze the effects of different urban permeability and building geometries on air pollutant dispersion in the street canyon. This was given by:

$$\overline{c} = <c>/<c_0>, \tag{7.3}$$

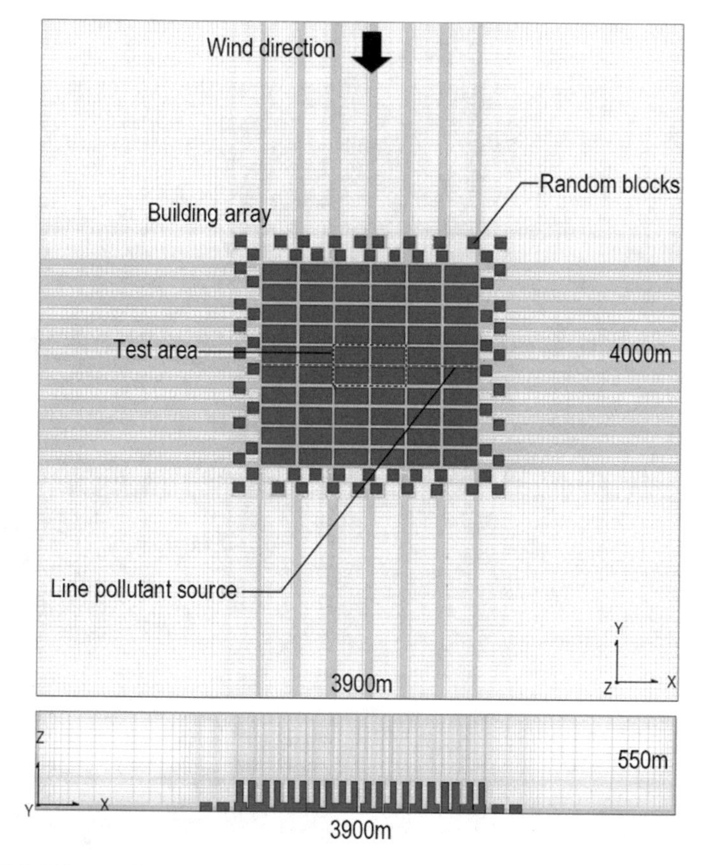

Fig. 7.6 Modeling configurations in the parametric study. Surrounding random blocks represent the surrounding urban surface roughness

where $<c>$ is the modeling result of the time-averaged concentration of NO_2 and $<c_0>$ is the reference emission concentration, the norm of $<c>$. Given the definition of $\overline{c}$, the threshold value of $\overline{c}$ was set to 1.0. When the value of $\overline{c}$ is less than 1.0, air pollutants start to disperse and do not concentrate in the street canyon.

7.4.1 Cross-Comparison Based on Normalized Concentration Contours

The contours of normalized concentration on horizontal planes (2 m above the ground) and vertical planes of eight cases are shown in Figs. 7.7 and 7.8, respectively, to give an intuitive grasp of the sensitivity of air pollutant dispersion to the changes in building geometries and permeability.

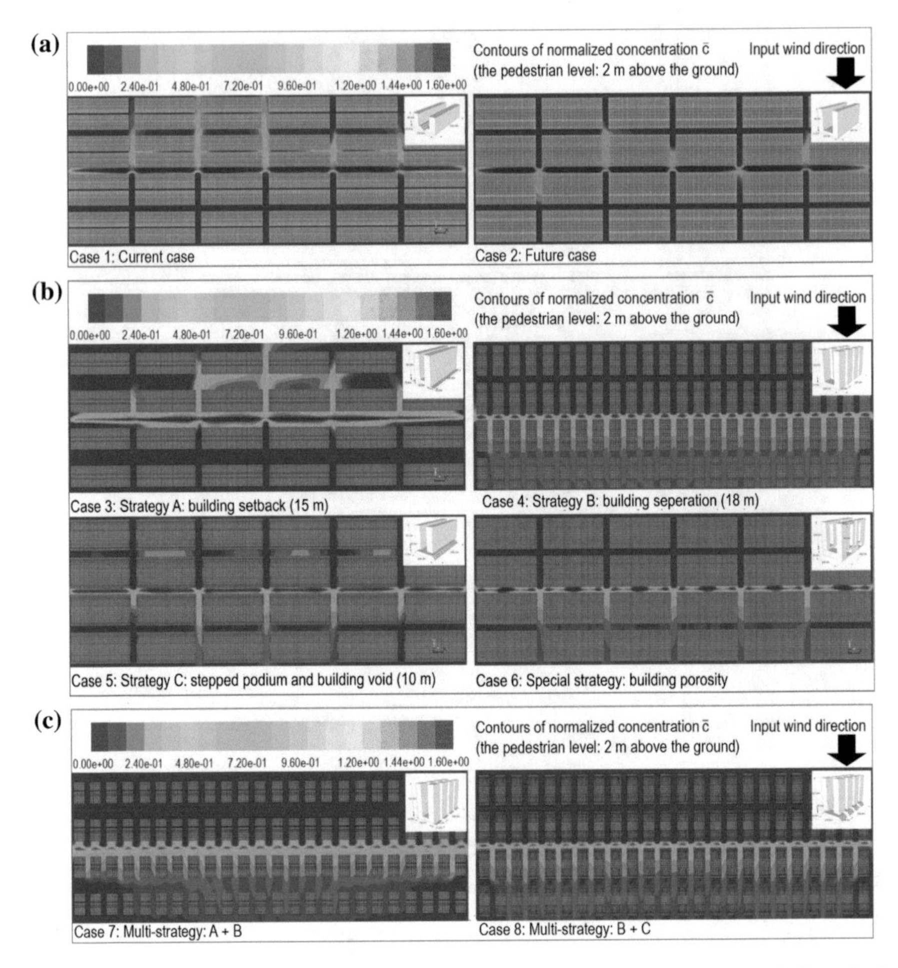

Fig. 7.7 Contours of $\overline{c}$ at the pedestrian level (2 m above the ground) in **a** Cases 1 and 2; **b** Cases 3–6 with single mitigation strategies; **c** Cases 7 and 8 with multi-mitigation strategies (For interpretation of the references to color in the text, the reader is referred to the electronic version of this book)

As shown in Fig. 7.7a, it is evident that the values of $\overline{c}$ in street canyons with emission sources were high in Case 1, the current case. Most values of pedestrian-level normalized concentration in streets with emission sources were larger than the threshold value (1.0), indicating that air pollutant was not dispersed but was concentrated at the pedestrian height levels. The vertical distributions shown in Fig. 7.8 indicated that the condition may worsen in the future, Case 2. The values of $\overline{c}$ in Case 2 were larger than 1.5 at the entire podium layer (0–15 m).

On the other hand, the mitigation strategies in Cases 3–8 can, in various degrees, promote air pollutant dispersion, as shown in Fig. 7.7b, c. Overall, the stagnant areas in Cases 3–8 are smaller than Cases 1 and 2, even though Cases 3–8 had higher plot ratios than Case 1, and the same plot ratio as Case 2 (Table 7.2). It should be noticed

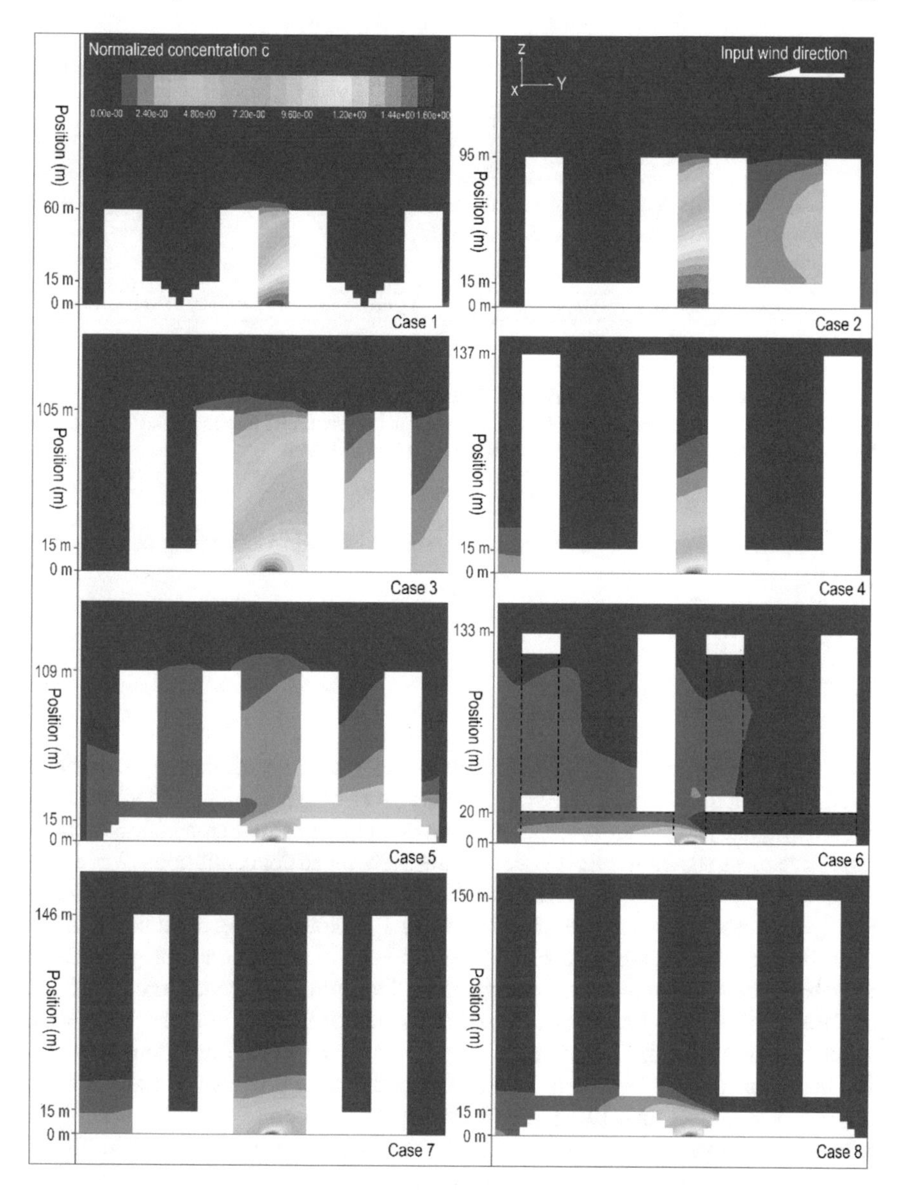

Fig. 7.8 Contours of $\overline{c}$ at the vertical plane in total eight cases (For interpretation of the references to color in the text, the reader is referred to the electronic version of this book)

that dispersion paths between Cases 3–8 were significantly different. In Cases 4–8, most of air pollutant was diluted by horizontal airflows through building porosities as shown in Fig. 7.8, so that only a small amount was diffused away from the top

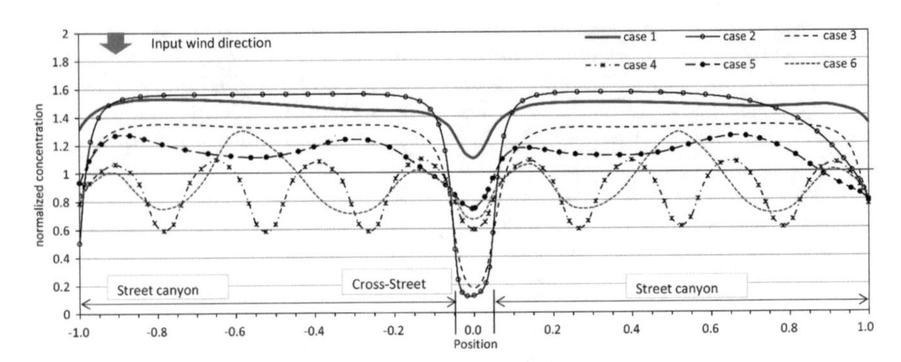

Fig. 7.9 Cross-comparison of the horizontal distributions of normalized concentration ($\overline{c}$) in Cases 1–6. The normalized concentration in Case 1 is plotted using a red solid line as a baseline case. The threshold value of the normalized concentration of 1.0 is highlighted

of canyon. By contrast, air pollutant was mainly diffused from the top of canyon in Case 3.

7.4.2 *Statistical Analysis*

A statistical analysis was conducted to evaluate the impact of various design strategies on air pollutant dispersion. Data were collected from both horizontal and vertical test lines to evaluate pedestrian-level outdoor and indoor air quality in roadside buildings, respectively. As shown in Fig. 7.9, data of the normalized concentrations (2 m above the ground) collected at the horizontal line at the pedestrian area are overlaid. In Fig. 7.9, the normalized concentration in Case 1 is plotted using a red solid line as a baseline case to cross-compare the cases with single mitigation strategies (Cases 3–6), with the highlighted threshold value of the normalized concentration of 1.0. In Case 3, air pollutant can be dispersed at the streets parallel with the input wind direction, but not in the target street canyon (i.e., $\overline{c}$ decreased by about 0.1–0.2). The performances of Cases 4–6 were significantly better than that of Case 3. Most values of $\overline{c}$ were less than 1.0 in Case 4. The stepped podium void in Case 5 did not promote air pollutant dispersion as much as in Case 4. Most values of $\overline{c}$ in Case 5 were slightly above 1.0. But an approximately 0.4 decrease of $\overline{c}$ was still observed. The performance of Case 6 was better than Case 5 and similar to Case 4, as most values of $\overline{c}$ were less than 1.0.

Multiple design strategies in Cases 7 and 8 were compared with Case 4 in Fig. 7.10 to clarify the effectiveness of combining single strategies. The performance of Case 7 was similar to Case 8, but only fractionally better than Case 4 (about 0.1 of $\overline{c}$ values). The values of $\overline{c}$ for the entire street were less than 1.0 in both Cases 7 and 8.

The vertical distributions of normalized concentration at the center of the building gap (vertical test line) are plotted in Fig. 7.11. Results indicated that the values of

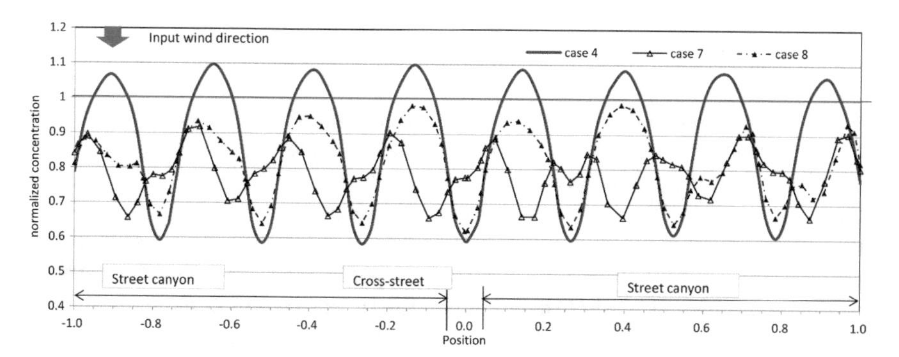

Fig. 7.10 Cross-comparison of the horizontal distributions of normalized concentration ($\overline{c}$) in Cases 4, 7, and 8

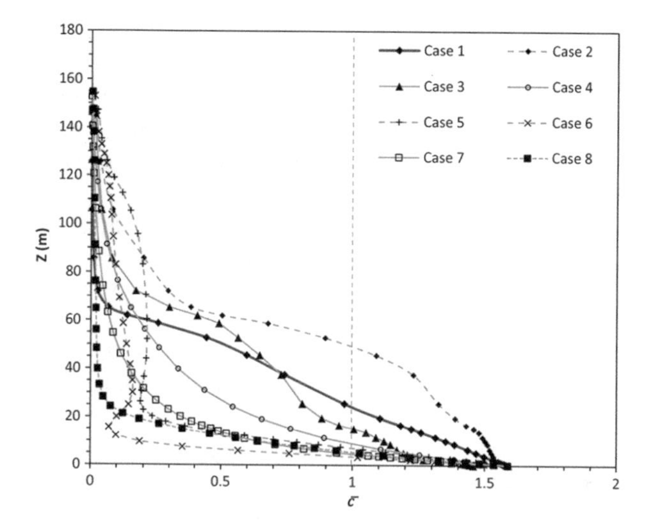

Fig. 7.11 Cross-comparison of the vertical distributions of normalized concentration ($\overline{c}$) in Cases 1–8

$\overline{c}$ were larger than 1.0 below 30 and 60 m respectively in Cases 1 and 2, and that deterioration in indoor air quality could occur in the roadside buildings at both height ranges due to outside traffic air pollutants, particularly for the residential building that typically adopts natural ventilation.

As shown in Fig. 7.11, the building setback in Case 3 did not mitigate this negative impact effectively, while the other design strategies were successful in reducing air pollutant concentrations. High concentration was only observed at 0–10 m height. With these design strategies, roadside buildings can enjoy natural ventilation and will not suffer from outside traffic air pollutants. This result is particularly important for urban design with residential land use.

7.5 Discussion

The above analyses indicate that the impacts of various mitigation strategies on pollutant dispersion in street canyons are similar to the ones on pedestrian-level natural ventilation (as discussed in Chap. 5), with a key difference that air pollutant dispersion significantly depends on the permeability of the entire street canyon layer. Interestingly, the incorporation of building porosity in Case 6, which was shown to be a poor mitigation strategy for improving the performance of pedestrian-level air ventilation in previous study (Yuan and Ng 2012), was proved in the current study to be effective in improving air pollutant dispersion both horizontally and vertically.

The vertical distribution of the normalized concentration ($\overline{c}$), turbulent intensity (I), and wind speed (U) at the vertical test line in the center of the building gap are plotted in Fig. 7.12 to further describe air pollutant dispersion in the street canyon and explain the cross-comparison results.

The analysis indicated that turbulent diffusion played a major role in pollutant dispersion in cases with low permeability (Cases 1–3: $P = 0.9$), and the pollutant dispersion significantly depends on convection effects in high permeability cases (Cases 4–6: $P = 0.7$–0.8). In Cases 1–3, the value of $\overline{c}$ in the street canyon decreased constantly with height. Regression analysis indicated that the vertical distribution of $\overline{c}$ in the low permeability cases (Cases 1–3) significantly depended on the turbulent intensity (I) ($R^2 = 0.87$), but no relationship between the normalized concentration and turbulent intensity ($R^2 = 0.11$) was observed in high permeability cases (Cases 4–6), as shown in Fig. 7.13.

In cases of higher permeability, building separation, stepped podium void, and building porosity all increased the effect of convection on pollutant dispersion. The normalized concentration did not depend on turbulent intensity, as shown in Fig. 7.13, nor decrease constantly with height. As shown in Fig. 7.11, the normalized concentration rapidly decreased until it reached the height of building permeability, and then slowly developed into an asymptotic value. In Cases 5 and 6, even though, the normalized concentration remained low despite the wind speed and turbulence intensity greatly decreased above the height of the building permeability. It was because the building permeability enabled clean air to flow horizontally into the street canyon, thereby diluting the pollutant concentrations and transporting them out of the street canyon. The building permeability in Case 4 was achieved by building separations that ranged from the ground level to the top of the building. Therefore, the strategy in Case 4 was better than in other cases. The building porosity in Case 6 provided more building permeability than Case 5, so that the pollutant dispersion in Case 6 was better than Case 5. The above analysis indicated that strategies for promoting convection effects, such as building porosity, separation and podium void, are more efficient to promote air pollutant dispersion than the ones for larger turbulent diffusion such as building setback.

Moreover, further examination of the eight contours of the normalized concentration at the pedestrian level Fig. 7.7 indicated that the direction of dispersion of pedestrian-level air pollutants depends on both input wind direction and the urban permeability. In Cases 1–3 ($P = 0.9$) with low permeability, the pedestrian-level air flow was reversed that was contrary to the input wind direction, so that air pollutants were transported upwind of the emission source. In contrast, in Cases 4–8 with high urban permeability ($0.6 \leq P \leq 0.8$), air pollutants were mainly dispersed downwind of the emission source. Therefore, unlike the understandings for low-density cities, estimation of the direction of pollutant dispersion at the pedestrian-level in high-density cities cannot depend solely on the prevailing wind direction. The urban permeability should also be taken into consideration for a reliable estimation.

7.6 Urban Permeability

Linear regression analysis was conducted in this study to statistically weight the effects of P and λ_p on the spatially averaged pollutant concentration. The results are plotted together in Fig. 7.14 that shows that the spatially averaged normalized concentration depended on the permeability of buildings (P), more than on the site coverage ratio (λ_p), as R^2 for P was 0.78, which was larger than R^2 (0.47) for λ_p.

Furthermore, to better predict the spatially averaged $\overline{c}$ in urban areas, integrated permeability λ_i was calculated based on Counihan's roughness model (Grimmond and Oke 1999) for estimating urban permeability as follows:

$$\lambda_i = P \cdot (C_1 \lambda_P + C_2) \quad (7.4)$$

The coefficients C_1 (1.4352) and C_2 (0.0463) (Grimmond and Oke 1999) were for the contribution of λ_p, considering that λ_p, as the horizontal permeability, was less important than P, as the vertical permeability. The values of λ_i in all eight cases were given in Table 7.2.

As shown in Fig. 7.14, a strong relationship between λ_i and spatially averaged $\overline{c}$ ($R^2 = 0.83$) indicated that λ_i was a good urban permeability index to estimate traffic air pollutant dispersion. This understanding provided a possibility for mapping traffic air pollutant concentrations in urban areas with traffic volume data by using GIS technology.

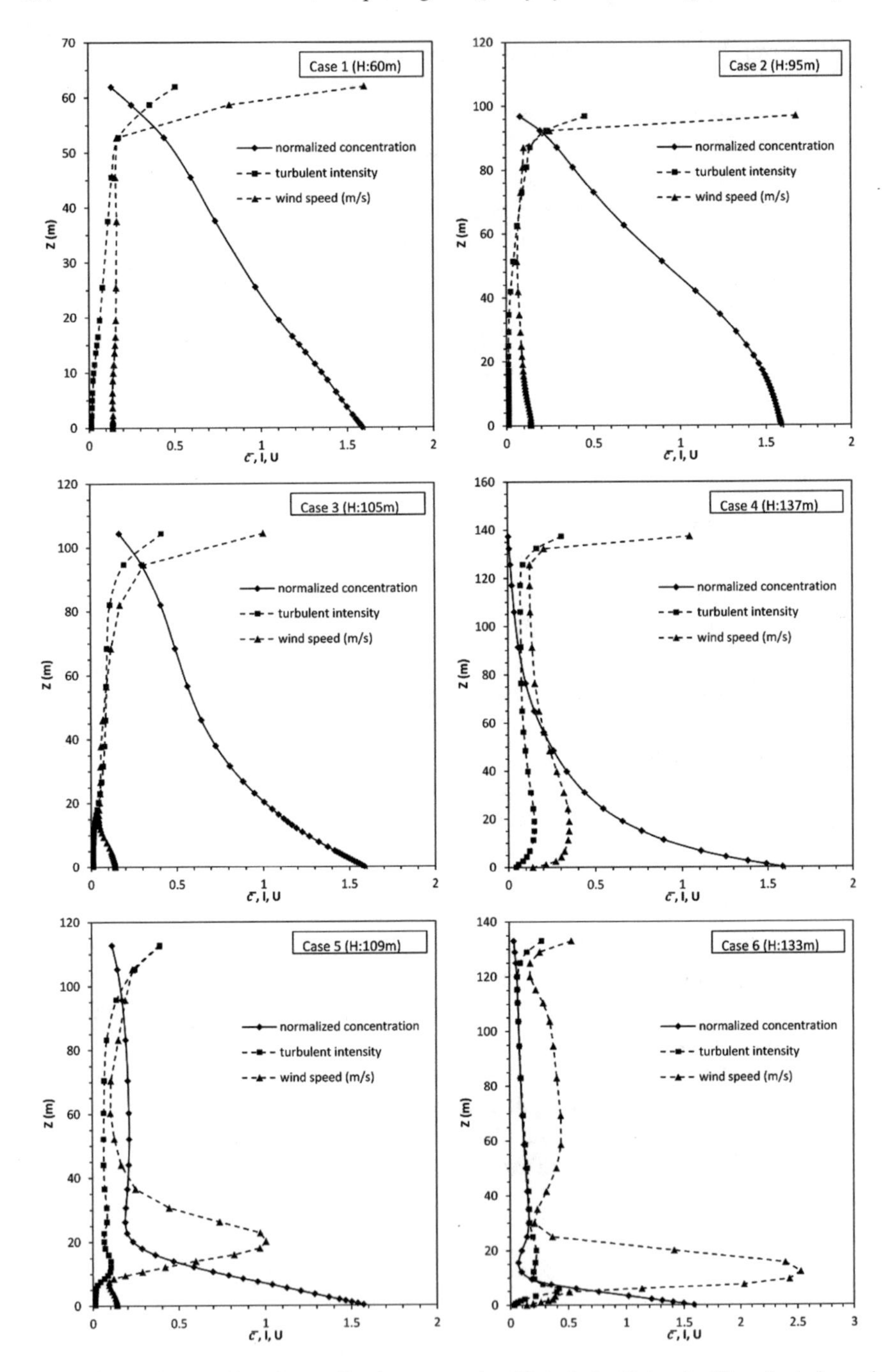

Fig. 7.12 Vertical profiles of normalized concentration ($\overline{c}$), turbulent intensity (I), and wind speed (U) in Cases 1–6. Building heights (H) in six cases are given

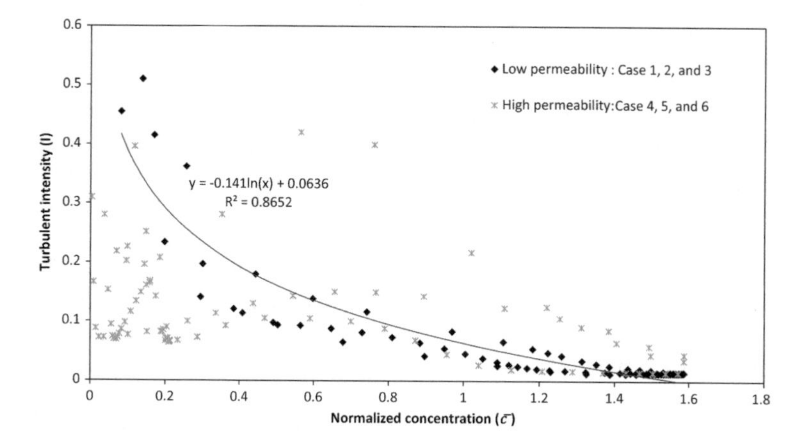

Fig. 7.13 Regression analysis of the relationship between turbulent intensity and normalized concentration in the street canyon at different cases with low and high permeability (*Significant level* 95%)

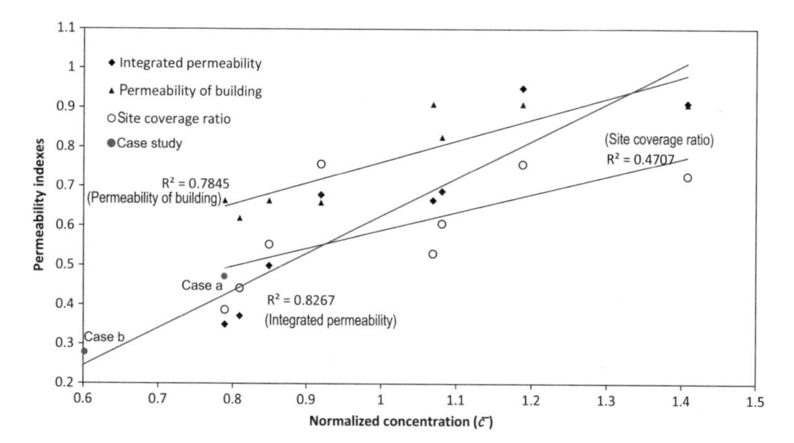

Fig. 7.14 Linear relationships between pedestrian-level air pollutant concentration and permeability indexes, P, λ_p, and λ_i (*Significant level* 95%)

7.7 Implementation in Urban Design

A case study at Mong Kok was conducted to demonstrate that the suggested design principles and understandings were feasible and applicable in an actual urban design practice. As shown in Fig. 7.15, an urban design (right, Case b) was produced from the current urban morphology (left, Case a) to improve the air quality at this high-density urban area. Based on the characteristic of different buildings, different mitigation design strategies were employed in every street block. To avoid reducing the land use efficiency, the plot ratio in these two cases remained the same. The site coverage ratios (λ_p) in Cases a and b were respectively 0.51 and 0.42. The averaged building permeability (*P*) of the total area was estimated at 0–60 m above the ground (0.7 in Case a and 0.5 in Case b).

For comparison purpose, a CFD simulation study was conducted to model airflows and traffic air pollutant dispersion in the above two cases. The simulations were constructed in accordance with the methodology in the parametric study. Figure 7.15 shows that the wind permeability in the entire area significantly increased in Case b and subsequently increased the local dispersion in the simulation study. The normalized concentration data were collected at the target street, and cross-compared

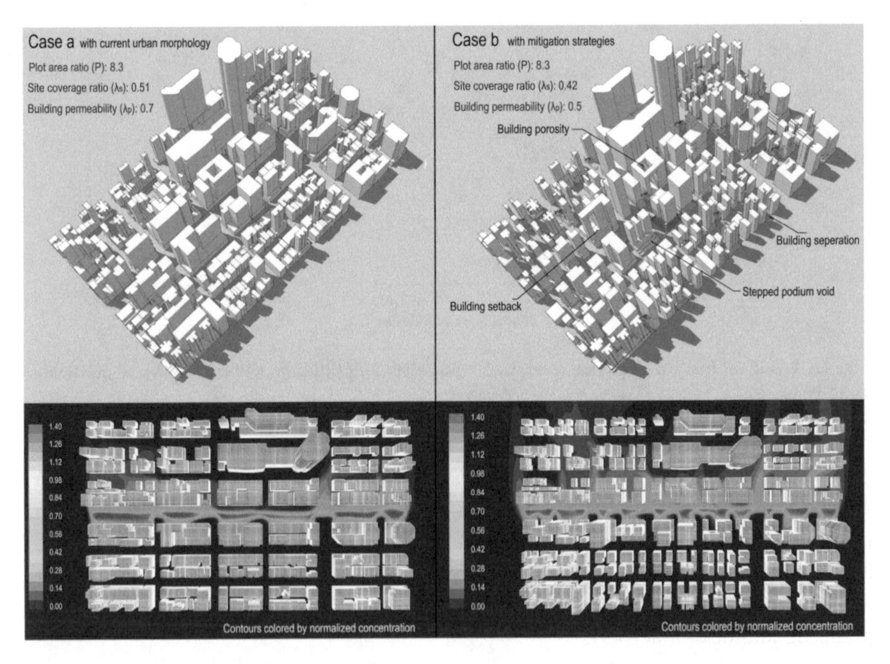

Fig. 7.15 Case study of urban redevelopment at Mong Kok and simulation results. Case a: the current urban area of Mong Kok; Case b: the urban morphology of Mong Kok with mitigation strategies. The mitigation strategies in Case b can significantly increase the wind permeability and decrease the traffic air pollutant concentration (For interpretation of the references to color in the text, the reader is referred to the electronic version of this book)

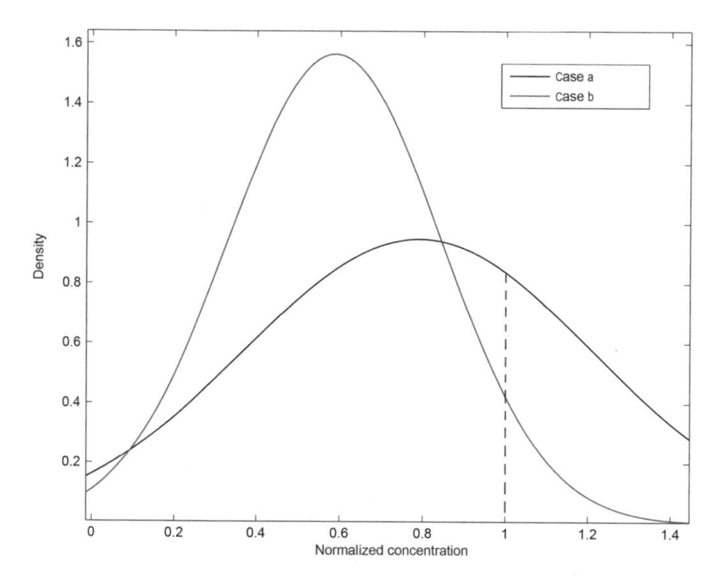

Fig. 7.16 Distribution of the normalized concentrations in Case a and Case b. The probability of high concentration ($\overline{c} > 1.0$) in Case b is far much less than the one in Case a

in Fig. 7.16. It shows that the probability of high concentration ($\overline{c} > 1.0$) in Case b was much less than the one in Case a. These results demonstrated the effectiveness of the suggested mitigation measures outlined in this study.

Based on the urban-scale understanding on permeability, the urban permeability λ_i was 0.48 in Case a and 0.27 in Case b, respectively. Similarly, the spatially averaged normalized concentration ($\overline{c}$) decreased from 0.79 in Case a to 0.60 in Case b. These results coincided well with the linear relationship shown in Fig. 7.14, and thus validated that the understanding of urban permeability can be feasible in the practical urban design.

7.8 Conclusion

This study demonstrated the use of CFD parametric approach to investigate the impact of urban permeability and building geometries on air pollutant dispersion in high-density urban areas. The analyses and discussion of the study results revealed the following scientific understandings for the decision-making in high-density urban planning and design activities:

(1) The SST κ–ω model can simulate the species transport in a street canyon with high accuracy. Thus, it can be considered as a good design tool in urban planning and architecture design for air pollutant problems because of its low computational cost and high accuracy. Cross-comparison of the LES and RANS

modeling methods with wind tunnel experiment results showed that the RANS model was sufficiently accurate for testing the impacts of planning and design activities on air pollutant concentration. Among the RANS models, the SST κ–ω model had the best performance because of the special near-wall region treatment using the standard κ–ω model.

(2) The simulation results supported the notion that mitigation strategies for the air pollutant dispersion are necessary in the planning and design of high-density urban areas. Traffic air pollutants are concentrated in the deep street canyon in current urban conditions, and these conditions will further deteriorate in the future urban development if the current planning and design activities go unchanged. Given the negative effects of air pollution on public health and of the high population density in Hong Kong, mitigation strategies are critical to alleviate this impact.

(3) The impacts of various urban planning and design strategies to improve air quality in both street canyons for pedestrian and inside adjacent buildings were quantitatively investigated and cross-compared in parametric studies. Upon gaining a better understanding of the mitigation strategies in particular planning and design cases, policymakers can be more confident in choosing the appropriate strategies to mitigate the negative effects of air pollutant on the surrounding environments. Strategies recommended in this study can be applied to both the new project design and urban redevelopment. The advantage of these kinds of building-scale strategies is that they allow the urban redevelopment to be done stepwise, one building by one building. It is more practical than the redevelopment done all at one time.

This study demonstrated that the air pollutant dispersion in high-density cities can be improved if strategies that promote convection effects (e.g., building separation, porosity, and stepped podium void) are implemented. Such strategies could be more efficient than strategies for larger turbulence diffusion, such as building setback. Contrary to AVA understanding, pedestrian-level pollutant concentration depends on the permeability of the entire street canyon. Although a high building porosity at the high level cannot increase the wind speed at pedestrian level, it can decrease pedestrian-level air pollutant concentrations. Therefore, appropriate strategies should be based on the particular concerns in different projects in the design and planning process.

(4) Simulation results indicated that both the prevailing wind direction and urban permeability are important to estimate the direction of pedestrian level pollutant dispersion in high-density cities. Low urban permeability in high-density urban areas could reverse air flow near the ground, allowing air pollutants to disperse into the windward area of the pollutant sources.

(5) By integrating the horizontal and vertical permeability based on Counihan's surface roughness model, a new permeability index λ_i was introduced to estimate the spatially averaged $\overline{c}$ at the pedestrian area. The efficiency of this index was validated by a linear regression analysis. This result provided a possibility for mapping air quality in the urban areas with traffic volume data by using GIS technology.

References

Ashie Y, Hirano K, Kono T (2009) Effects of sea breeze on thermal environment as a measure against Tokyo's urban heat island. Paper presented at the seventh international conference on urban climate, Yokohama, Japan

Baik JJ, Kang YS, Kim JJ (2007) Modeling reactive pollutant dispersion in an urban street canyon. Atmos Environ 41(5):934–949

Blocken B, Stathopoulos T, Carmeliet J (2007) CFD simulation of the atmospheric boundary layer: wall function problems. Atmos Environ 41(2):238–252

Buccolieri R, Sandberg M, Di Sabatino S (2010) City breathability and its link to pollutant concentration distribution within urban-like geometries. Atmos Environ 44(15):1894–1903

Eeftens M, Beekhuizen J, Beelen R, Wang M, Vermeulen R, Brunekreef B, Huss A, Hoek G (2013) Quantifying urban street configuration for improvements in air pollution models. Atmos Environ 72:1–9

European Environment Agency (EEA) (2012) Air quality in Europe—2012 report. Denmark, Copenhagen

Fluent Inc. (2012) FLUENT 14.0 theory Guide

Givoni B (1969) Man, climate and architecture. Elsevier, New York

Gousseau P, Blocken B, Stathopoulos T, van Heijst GJF (2010) CFD simulation of near-field pollutant dispersion on a high-resolution grid: a case study by LES and RANS for a building group in downtown Montreal. Atmos Environ 45(2):428–438

Gousseau P, Blocken B, van Heijst GJF (2011) CFD simulation of pollutant dispersion around isolated buildings: on the role of convective and turbulent mass fluxes in the prediction accuracy. J Hazard Mater 194:422–434

Grice S, Stedman J, Kent A, Hobson M, Norris J, Abbott J, Cooke S (2009) Recent trends and projections of primary NO_2 emissions in Europe. Atmos Environ 43(13):2154–2167

Grimmond CSB, Oke TR (1999) Aerodynamic properties of urban areas derived from analysis of surface form. J Appl Meteorol 38:1262–1292

Hang J, Li YG, Sandberg M, Buccolieri R, Di Sabatino S (2012) The influence of building height variability on pollutant dispersion and pedestrian ventilation in idealized high-rise urban areas. Build Environ 56:346–360

Ho KF, Ho SSH, Lee SC, Cheng Y, Chow JC, Watson JG, Louie PKK, Tian LW (2009) Emissions of gas- and particle-phase polycyclic aromatic hydrocarbons (PAHs) in the Shing Mun Tunnel, Hong Kong. Atmos Environ 43(40):6343–6351

Hong Kong Building Department (HKBD) (2006) Sustainable building design guidelines. Practical note for authorized persons, registered structure engineers and registered geotechnical engineers. APP-152, Hong Kong

Hong Kong Environmental Protection Department (HK EPD) (2011) Air quality in Hong Kong 2011—a report on the results from the air quality monitoring network (AQMN). The Goveronment of the Hong Kong Special Administrative Region, Hong Kong

Huang H, Ooka R, Chen H, Kato S, Takahashi T, Watanabe T (2008) CFD analysis on traffic-induced air pollutant dispersion under non-isothermal condition in a complex urban area in winter. J Wind Eng Ind Aerodyn 96:1774–1788

Kikumoto H, Ooka R (2012) A numerical study of air pollutant dispersion with bimolecular chemical reactions in an urban street canyon using large-eddy simulation. Atmos Environ 54:456–464

Kondo H, Asahi K, Tomizuka T, Suzuki M (2006) Numerical analysis of diffusion around a suspended expressway by a multi-scale CFD model. Atmos Environ 40(16):2852–2859

Kristensson A, Johansson C, Westerholm R, Swietlicki E, Gidhagen L, Wideqvist U, Vesely V (2004) Real-world traffic emission factors of gases and particles measured in a road tunnel in Stockholm, Sweden. Atmos Environ 38(5):657–673

Kunzli N, Kaiser R, Medina S, Studnicka M, Chanel O, Filliger P, Herry M, Horak F, Puybonnieux-Texier V, Quenel P, Schneider J, Seethaler R, Vergnaud JC, Sommer H (2000) Public-health impact of outdoor and traffic-related air pollution: a European assessment. Lancet 356(9232):795–801

Li XX, Liu CH, Leung DYC, Lam KM (2006) Recent progress in CFD modelling of wind field and pollutant transport in street canyons. Atmos Environ 40(29):5640–5658

Menter FR, Kuntz M, Langtry R (2003) Ten years of industrial experience with the SST turbulence model. Turbul Heat Mass Transf 4:625–632

Mirzaei PA, Haghighat F (2012) A procedure to quantify the impact of mitigation techniques on the urban ventilation. Build Environ 47:410–420

Mochida A, Murakami S, Ojima T, Kim S, Ooka R, Sugiyama H (1997) CFD analysis of mesoscale climate in the Greater Tokyo area. J Wind Eng Ind Aerodyn 67–68:459–477

Ng E (2012) Towards a planning and practical understanding for the need of meteorological and climatic information for the design of high density cities—a case based study of Hong Kong. Int J Climatol 32:582–598

Ng E, Wong NH, Han M (2006) Parametric studies of urban design morphologies and their implied environmental performance. In: Bay JH, Ong BL (ed). Architectural Press, London

Ng E, Yuan C, Chen L, Ren C, Fung JCH (2011) Improving the wind environment in high-density cities by understanding urban morphology and surface roughness: a study in Hong Kong. Landscape Urban Plann 101(1):59–74

Pontiggia M, Derudi M, Alba M, Scaioni M, Rota R (2010) Hazardous gas releases in urban areas: assessment of consequences through CFD modelling. J Hazard Mater 176(1–3):589–596

Richmond-Bryant J, Reff A (2012) Air pollution retention within a complex of urban street canyons: a two-city comparison. Atmos Environ 49:24–32

Salim SM, Buccolieri R, Chan A, Di Sabatino S (2011) Numerical simulation of atmospheric pollutant dispersion in an urban street canyon: comparison between RANS and LES. J Wind Eng Ind Aerodyn 99(2–3):103–113

Stathopoulos T, Storms R (1986) Wind environmental conditions in passages between buildings. J Wind Eng Ind Aerodyn 95:941–962

Tominaga Y, Stathopoulos T (2011) CFD modeling of pollution dispersion in a street canyon: comparison between LES and RANS. J Wind Eng Ind Aerodyn 99(4):340–348

Tominaga Y, Stathopoulos T (2012) CFD modeling of pollution dispersion in building array: evaluation of turbulent scalar flux modeling in RANS model using LES results. J Wind Eng Ind Aerodym 104:484–491

Tominaga Y, Mochida A, Murakami S, Sawaki S (2008a) Comparison of various revised k-epsilon models and LES applied to flow around a high-rise building model with 1:1:2 shape placed within the surface boundary layer. J Wind Eng Ind Aerodyn 96(4):389–411

Tominaga Y, Mochida A, Yoshie R, Kataoka H, Nozu T, Yoshikawa M, Shirasawa T (2008b) AIJ guidelines for practical applications of CFD to pedestrian wind environment around buildings. J Wind Eng Ind Aerodyn 96:1749–1761

Vardoulakis S, Fisher BEA, Pericleous K, Gonzalez-Flesca N (2003) Modelling air quality in street canyons: a review. Atmos Environ 37(2):155–182

Wang M, Lin CH, Chen QY (2012) Advanced turbulence models for predicting particle transport in enclosed environments. Build Environ 47:40–49

World Health Organization (WHO) (2008) Air quality and health—Fact sheet no. 313. Accessed 8 July 2012

Yim SHL, Fung JCH, Lau AKH (2009a) Mesoscale simulation of year-to-year variation of wind power potential over southern China. Energies 2:340–361

Yim SHL, Fung JCH, Lau AKH, Kot SC (2009b) Air ventilation impacts of the "wall effect" resulting from the alignment of high-rise buildings. Atmos Environ 43(32):4982–4994

Yuan C, Ng E (2012) Building porosity for better urban ventilation in high-density cities—a computational parametric study. Build Environ 50:176–189

Zhang Z, Chen Q (2006) Experimental measurements and numerical simulations of particle transport and distribution in ventilated rooms. Atmos Environ 40(18):3396–3408

Chapter 8
A Semi-Empirical Model for Urban Trees Effects on the Wind Environment

8.1 Introduction

In the context of rapid urbanization and depletion of natural resources, high-density urban living for better allocation of natural resources is an inevitable growing trend. However, such high-density nature also has negative impacts on urban living, such as stacked housing and crowded living with poor air quality and high thermal discomfort. As one of the essential elements in urban areas, urban vegetation can improve thermal comfort (Dimoudi and Nikolopoulou 2003; Ng et al. 2012), enhance psychological health of urban dwellers (Thompson et al. 2014), and promote the urban biodiversity (Savard et al. 2000; Kowarik 2011). As many initiatives of adding urban vegetation were introduced into metropolitan cities, landscape planning has become critical in the whole urban planning system especially in (sub) tropical cities. For example, the average per capita green space provision is 10 m^2 in Singapore (Singapore National Parks Board 2015), 7 m^2 in Tokyo (Tokyo Metropolitan Government 2007), 2 m^2 in buildable land areas in Hong Kong with 40% of land designated as the green nature reserve (Hong Kong Planning Department (HKPD) 2011), and 12.5 m^2 in Shanghai (Bureau of Shanghai World Expo Coordination 2009).

As shown in Fig. 8.1, most vegetation, as defined by the Normalized Difference Vegetation Index (NDVI) distribution in Hong Kong, is located at country parks, while the greenery (e.g., trees) in urban areas is very limited. Since urban trees can benefit the urban living by directly influencing the microclimate in the street canyon (Ng et al. 2012), the Hong Kong Civil Engineering and Development Department (2010) conducted the detailed Greening Master Plan (GMP) to introduce more vegetation into urban areas. As shown in Fig. 8.2, several GMPs have been implemented

---

Originally published in Yuan Chao, Leslie K. Norford and Edward Ng, 2017. A semi-empirical model for the effect of trees on the urban wind environment. Landscape and Urban Planning, 168, pp. 84–93, © Elsevier, https://doi.org/10.1016/j.landurbplan.2017.09.029

C. Yuan, *Urban Wind Environment*, SpringerBriefs in Architectural Design and Technology, https://doi.org/10.1007/978-981-10-5451-8_8

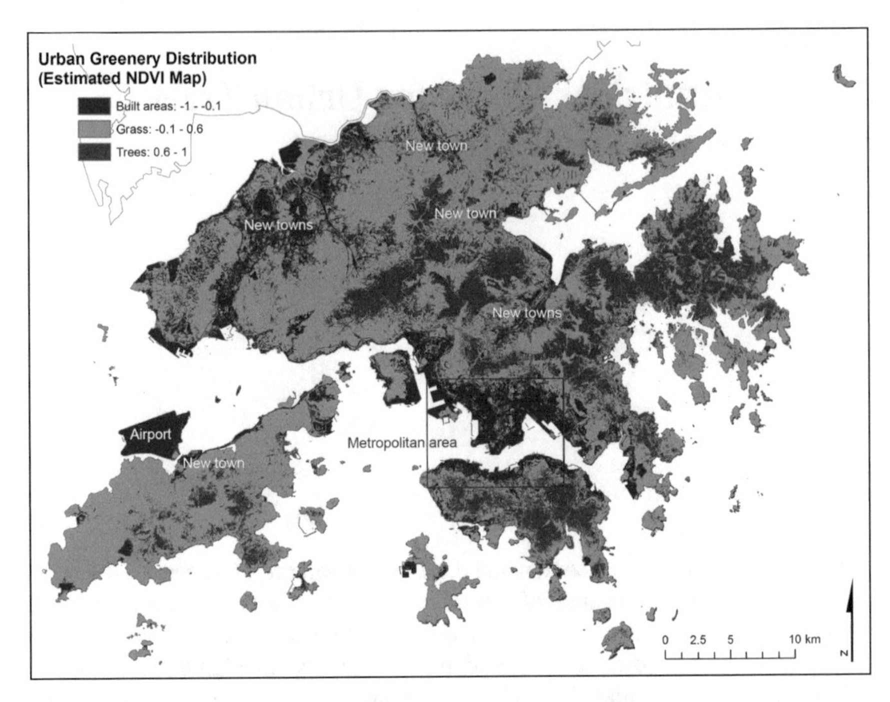

Fig. 8.1 NDVI distribution at Hong Kong. Most of urban vegetation is located at country parks, and limited greenery are located at the urban areas (For interpretation of the references to color in the text, the reader is referred to the electronic version of this book)

since 2004 in the highest density districts, such as Sheung Wan, Mong Kok, and Tsim Sha Tsui. However, as the potential areas for planting trees in urban areas remain very limited in GMP, it is difficult to increase the planting area in high-density urban areas. As shown in Fig. 8.3, there is no significant increment on the proposed greenery ratio resulting from either the short- or long-term GMP. On the other hand, in areas with increased vegetation, the negative effects of urban trees (e.g., stagnated airflow, trapped air pollutant, and anthropogenic heat) on urban environment should not be ignored (Vos et al. 2013; Gromke and Ruck 2008). Therefore, a quantitative and more detailed study is needed to understand the environmental functions of urban trees in the street canyon, and maximize their benefits to the urban environment.

Trees have three major functions in the urban microclimate: they (1) provide shading to reduce heat gain on buildings and ground (Shahidan et al. 2010); (2) transpire water to atmosphere to decrease heat storage in the urban canopy layer (Loughner et al. 2012), and (3) resist wind (Cullen 2005; Gromke and Ruck 2008). The shading and evaporation aspects of trees have been well documented, and are considered as positive impacts to the urban environment, because they can either improve ambient thermal comfort by directly blocking the solar radiation or mitigate the urban heat island by removing heat from the urban canopy layer. However, the wind resistance aspect of trees is still not fully understood. Conventional studies have

Fig. 8.2 Green master planning (GMP) in Tsim Sha Tsui (above), and the street canyon after the GMP at Sheung Wan, Mong Kok, and Tsim Sha Tsui (below) (Hong Kong Civil Engineering and Development Department 2010)

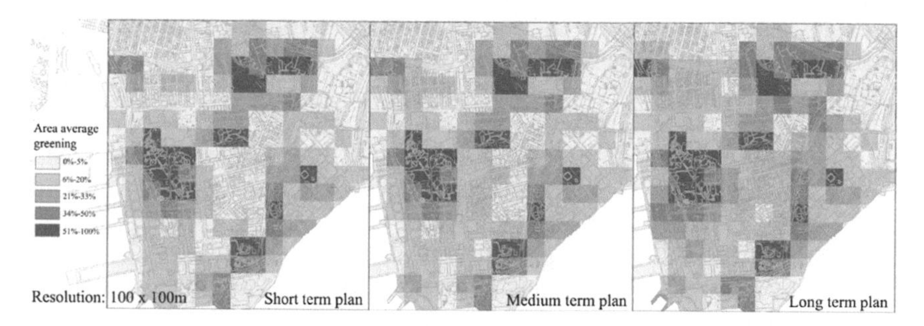

Fig. 8.3 Proposed greenery ratio at Tsim Sha Tsui in the short, medium, and long term

mainly focused on pedestrian safety and tree risk, e.g., breakage of large roadside trees as a result of strong gusts (Cullen 2005). In high-density urban areas, wind speed tends to be slower. As the mean wind speed at 20 m above the ground level in urban area of Hong Kong decreased by about 40%, from 2.5 to 1.5 m/s, over the past decade (HKPD 2005), outdoor natural ventilation becomes an environmental issue. Consequently, when there is limited outdoor natural ventilation and a growing need to introduce more vegetation into urban areas, it is not only important to focus on the drag force of buildings on airflow but is also critical to study the additional drag force of trees, in order to address outdoor natural ventilation issues in the urban landscape planning.

Computational Fluid Dynamics (CFD) simulation and wind tunnel experiment are two major methods for investigating the tree effect on surrounding airflow (Gross 1987; Aubrun et al. 2005). However, it might be more difficult to model aerodynamic properties of trees due to porosity and flexibility of plants than modeling buildings in the CFD and wind tunnel experiment. In the city scale, Li and Norford (2016) conducted a Weather Research and Forecasting (WRF) modeling (300 × 300 m resolution) to investigate the cooling effect of vegetation in urban areas, but no aerodynamic effects of urban trees in the urban canopy layer were included. In the neighborhood and building scales, several studies have included both individual and group of trees in the CFD simulation and wind tunnel experiments to measure the aerodynamic drag force of trees (Mochida et al. 2008; Hiraoka 1993; Gromke and Ruck 2008). While these studies provide important knowledge on the aerodynamic properties of tress, they did not include buildings in the modeling in most cases. Even though sharp-edged buildings dominate airflow in the urban boundary layer, it is still necessary to clarify the aerodynamic effects of trees within the urban context. Gromke (2011) did just that by conducting the wind tunnel experiment to investigate the effect of roadside trees on the air pollutant dispersion in the typical street canyon, by measuring the pressure loss coefficient. This experiment was later reproduced by Jeanjean et al. (2015) using CFD simulation and then applied in the real urban scenario (East Midlands region of the UK). While previous studies provide some informative results, they required intensive technical support and high computational capability. Also, as these studies were mostly used for research purposes, the implementation of their research outputs to landscape planning practice remains limited. Therefore, a more practical modeling method is needed for the design and planning practice use.

Morphological method has been considered as a practical tool for such purpose (Grimmond and Oke 1999; Wong et al. 2010; Ng et al. 2011). This method correlates geometric indices, such as frontal area density (λ_f) and site coverage ratio (λ_p), with experimental wind data by statistical distribution fitting. The experimental wind data could be either the wind speed or the aerodynamic indices, such as roughness length (z_0) and displacement height (z_d). It has been shown that these wind data are strongly related to the geometric indices. Thus, the aerodynamic effect of buildings can be parameterized and modeled, and the complicated calculations of fluid mechanics can be avoided, significantly lowering the computational costs. These geometric indices calculated by pixels in a Geographic Information System (GIS) have been used to evaluate wind environment in spatially continuous heterogeneity urban areas (Gál and Unger 2009; Ng et al. 2011; Wong et al. 2010; Yim et al. 2009; Yuan et al. 2014, 2016). Nonetheless, there are currently no existing morphological models suitable for general use, given that the development of these models is mainly based on the statistical fitting, and most existing models do not include trees in the street canyon (Grimmond et al. 1998).

Compared with the morphological method, the algorithms with urban geometric indices in semiempirical urban canopy models are derived from physical understandings, such as the balance between momentum flux and drag force, rather than statistical fitting (Lettau 1969; MacDonald et al. 1998; Coceal and Belcher 2004). On the other hand, the model coefficients (e.g., drag force coefficient) are deduced from the experiment or field measurement data. Therefore, the urban canopy models are semiempirical. Bentham and Britter (2003) developed a practical and comprehensive urban canopy model to estimate the averaged wind speed (U_c) at the urban canopy layer, and MacDonald et al. (1998) conducted an urban canopy model to estimate the vertical wind profile. Despite that, most of the existing urban canopy models only include buildings, but not urban trees, given that buildings dominate the microenvironment in the urban boundary layer as mentioned earlier.

The study described in this chapter aimed to develop a new urban canopy model to correlate the urban density and tree species indices with wind speed in the street canyon, so that the new urban canopy model could help informing decision-making more efficiently during landscape planning. By coordinating the information on momentum flux and porosity of trees from the literature review, a new urban canopy model was developed based on the balance between the momentum flux from the upper layer and drag force of buildings and urban trees on airflow. The momentum flux and porosity of tropical urban trees were parametrized respectively. Both urban density and tree species indices were then inserted into the new semiempirical model to calculate the wind speed in the street canyon. In doing so, we identified key ways to reduce the negative impact of urban trees on the wind environment, and provided corresponding implementation.

8.2 Development of Modeling Method

8.2.1 Balance Between Momentum Flux and Drag Force

The development of the new modeling method was based on the balance between momentum flux and the drag force of both buildings and trees on airflow. It was assumed that the momentum flux above the urban canopy layer and the surface shear stress only depend on the buildings, as the buildings have larger drag force than trees, and they dominate airflow in the urban boundary layer (Krayenhoff et al. 2015). Given the steady and uniform airflow, the balance in the urban canopy layer between momentum flux (left side of Eq. 8.1) and drag force of both buildings and trees on airflow (right side of Eq. 8.1) can be stated as:

$$\tau_{\mathrm{w}} A_{\mathrm{site}} = \frac{1}{2}\rho U_{(\mathrm{c})}^2 / \left(1-\lambda_{\mathrm{p}}\right)\left(\sum\left(C_{\mathrm{D\_building}} A_{\mathrm{front\_building}}\right) + \sum\left(C_{\mathrm{D\_tree}} A_{\mathrm{front\_tree}}\right)\right), \tag{8.1}$$

where τ_{w} is the vertical flux of horizontal momentum from the upper layer to the lower layer due to the turbulence mixing effect, and can be expressed as $\tau_{\mathrm{w}} = \rho u_*^2$, in which u_* is the friction velocity. λ_{p} is the site coverage ratio, A_{front} is the frontal area, A_{site} is the site area, and ρ is the air density. The first and second items on the right side of equation are the drag force of buildings and trees, respectively. Some studies (Kitagawa et al. 2015; Gromke and Ruck 2008) indicated that the horizontal wind force on a tree varies linearly with increasing wind speed due to the decrease of trees' frontal area, and thus the velocity exponent in Eq. 8.1 is equal to 1, instead of 2. But it can only be applied if there is very strong wind, and for the tree risk management study. Even though there is no clear threshold value of wind speed suggested by existing researches, the conventional drag equation with velocity squared was still considered to be appropriate in this study, as the averaged wind speed in the urban canopy layer (UCL) is rarely large enough to reshape the plant crown. Therefore, the averaged wind speed (U_c) in the urban canopy layer normalized by the friction velocity, u_*, can be expressed as:

$$\frac{U_c}{u_*} = \left(\frac{2(1-\lambda_p)}{C_{D\_building}\lambda_{f\_building} + C_{D\_tree}\lambda_{f\_tree}}\right)^{0.5}, \quad \text{in which } \lambda_f = A_{front}/A_{site}, \tag{8.2}$$

To close Eq. 8.2 and solve U_c, we need to identify: (1) friction velocity (u_*); (2) drag coefficient of trees $C_{D(tree)}$ and buildings $C_{D(building)}$; and (3) frontal area density of buildings $(\lambda_{f\_building})$ and trees (λ_{f_tree}). Based on the literature, we chose $C_{D(building)}$ as 2.0 (Coceal and Belcher 2004) and $C_{D(tree)}$ as 0.8 that is appropriate for the low wind velocities (<10 m/s) (Gromke and Ruck 2008), respectively. $\lambda_{f\_building}$ was calculated as the sectional frontal area density, which can better represent effect of building on the airflow due to the skimming flow associated with high-density urban areas (Ng et al. 2011; Yuan et al. 2016). Therefore, in following sections, we parameterized friction velocity (u_*) and frontal area density of trees (λ_{f_tree}) as the unknown variables.

8.2.2 Parametrization of Friction Velocity (u_*)

Friction velocity (u_*) is directly related with surface shear stress, and defined as $\sqrt{\tau_w/\rho}$ in the fluid mechanics (Schlichting and Gersten 2000). With the unit of velocity, u_* is one of the basic variables to describe the near-surface flow (Britter and Hanna 2003). Compared with mean velocity (U_{ref}) at the reference height (z_{ref}), u_* does not depend on the boundary layer height (Schlichting and Gersten 2000), and thus u_* is frequently used to nondimensionalize other velocity variables, such as the use of u_* in the logarithmic law equation (Eq. 8.3). Except for the well-controlled wind tunnel experiments that can directly measure surface shear stress (τ_w) (Cheng and Castro 2002), u_* can be estimated by turbulence fluctuation using ultrasonic anemometer in the field (Walker 2005). But similar with wind speed measurement, a representative observation of turbulence fluctuation in urban areas is difficult (Oke 2006). While commercial LIDAR (Light Detection and Ranging)'s technology (airplane/satellite based) and Doppler Weather Radar (Ground based) have been used to detect the wind shear, for example at airport and wind form (Bot 2014; Kumer et al. 2016), they are rarely applied in urban areas with stagnant airflow.

$$\frac{U_{ref}}{u_*} = \frac{1}{\kappa} \ln \frac{z_{ref} - z_d}{z_0} \tag{8.3}$$

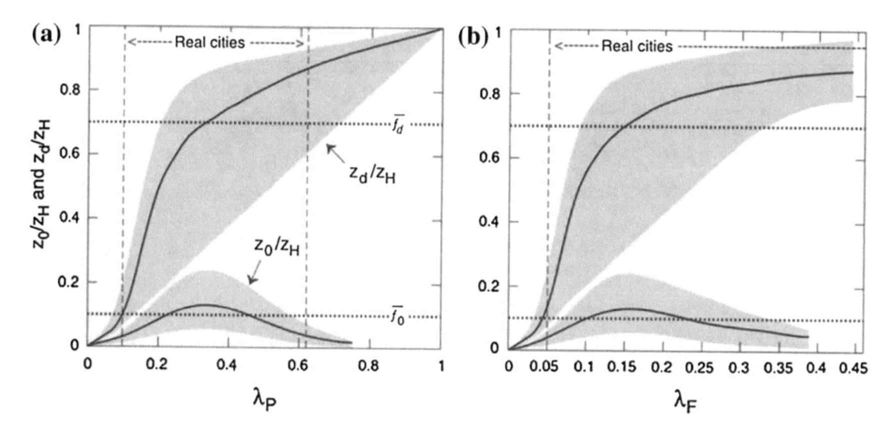

Fig. 8.4 Determination of z_0, z_d using λ_f and λ_p (Grimmond and Oke 1999)

We estimated u_* at urban areas using the logarithmic law as shown in Eq. 8.3. This method is not as straightforward as the wind tunnel experiment and field measurement, since it requires to model the roughnesslength z_0 and the displacement height z_d. But, this curve-fitting method uses the existing methods available to estimate z_0 and $z_d\lambda_f$ and thus is more practical than the wind tunnel experiment and field measurement. More importantly, this method makes it possible to clarify the relationship between u_* and the geometry of surface roughness elements. Grimmond and Oke (1999) conducted a broad and critical review on the existing morphological methods to model z_0 and z_d, and provided a general understanding of the relationship between z_0, z_d, λ_f, and λ_p as shown in Fig. 8.4. In the current study, we directly used these fitting curves, the solid lines in Fig. 8.4, to list the values of z_0 and z_d with corresponding values of λ_f. Because z_0 and z_d were normalized by z_h, we used the height of roughness sub-layer z_* as the reference height, and take z_* as $2z_h$ for the compact building canopies, as Roth (2007) and Raupach (1992) suggested (Table 8.1).

Subsequently, u_* normalized by the mean wind speed at the top of roughness sub-layer z_* (U_{ref}) can be estimated by λ_f as shown in Fig. 8.5. Similar with z_0, the sensitivity of u_* on increasing λ_f significantly decreases when λ_f is larger than 0.4, i.e., u_* is almost constant when λ_f is larger than 0.4. It is reasonable, since surface shear stress decreases when more roughness elements are included, and they start to interfere with each other, i.e., the skimming flow (Oke 1987). Quantitatively, u_*/U_{ref}

Table 8.1 Variation of z_0 and z_d with λ_f from 0.05 to 1.0 to estimate u_*. Total 11 cases were included

	0	1	2	3	4	5	6	7	8	9	10
λ_f	0.05	0.10	0.20	0.30	0.40	0.50	0.60	0.70	0.80	0.90	1.00
z_d/z_h	0.10	0.55	0.76	0.81	0.87	0.92	0.94	0.96	0.98	0.99	1.00
z_0/z_h	0.03	0.10	0.11	0.07	0.04	0.038	0.036	0.035	0.03	0.03	0.03

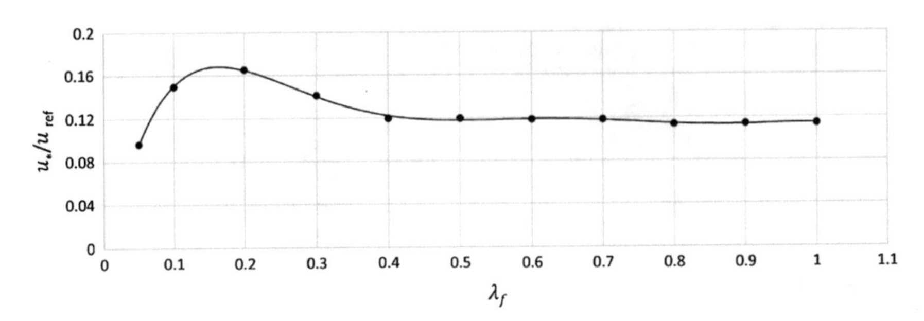

Fig. 8.5 Relationship between u_*/U_{ref} and λ_{f}. u_* is almost constant, 0.12, when λ_{f} is larger than 0.4

is a constant 0.12 with $\lambda_{\text{f}} \geq 0.4$, which covers most of urban areas, and u_* can be expressed as:

$$u_* = 0.12\,U_{\text{ref}}, \quad \text{when } \lambda_{\text{f\_building}} \geq 0.4, \tag{8.4}$$

To validate the modeling results of u_*, we compared them with the wind data from wind tunnel experiment and CFD simulations. The wind tunnel experiment was conducted by Hong Kong University of Science and Technology for the Air Ventilation Assessment (AVA) project in Hong Kong (HKPD 2008), whereas the CFD simulation data is derived from Architectural Institute of Japan (AIJ) for the cases of city blocks (Tominaga et al. 2008). With the values of u_*/U_{ref} in Fig. 8.5 as the input data, we used Bentham and Britter model (Bentham and Britter 2003) to calculate $U_{\text{c}}/U_{\text{ref}}$, as the solid line shown in Fig. 8.6. The observed data, i.e., AVA and AIJ data, were overlaid with modeling results for the cross-comparison. *p*-value of 0.75 in the hypothesis test was also obtained, which indicated that there was no significant difference between modeling results and measurement data (i.e., AVA and AIJ data) at the significance level of 0.05. Thus, this showed that the modeling method in this study can accurately estimate u_*, so that mean wind speed in the urban canopy layer can be easily calculated by λ_{f} as Eq. 8.5 (non-tree scenario) (Bentham and Britter 2003). For the scenario with both buildings and trees, Eq. 8.2 can be expressed as Eq. 8.6 where both building and tree drag force are included by parameterizing the friction velocity u_*.

$$\frac{U_{\text{C}}}{U_{\text{ref}}} = 0.12\left(\frac{\lambda_{\text{f}}}{2}\right)^{-0.5}, \quad \text{when } \lambda_{\text{f\_building}} \geq 0.4 \tag{8.5}$$

$$\frac{U_{\text{C}}}{U_{\text{ref}}} = 0.12\left(\frac{2(1-\lambda_{\text{p}})}{C_{\text{D\_building}}\lambda_{\text{f\_building}} + C_{\text{D\_tree}}\lambda_{\text{f\_tree}}}\right)^{0.5}, \quad \text{when } \lambda_{\text{f\_building}} \geq 0.4 \tag{8.6}$$

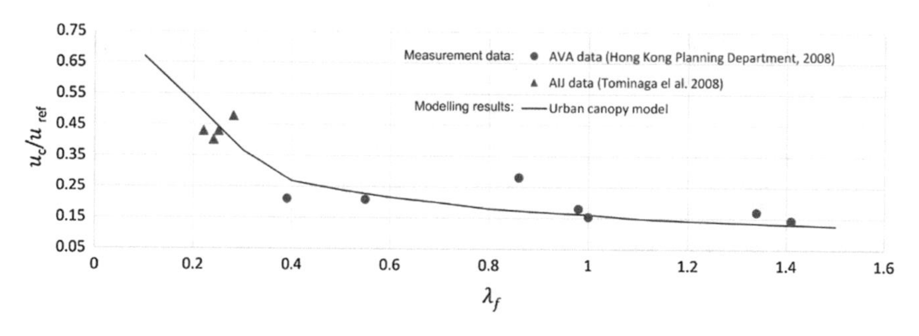

Fig. 8.6 Validation. The new urban canopy model matches the experimental data well (*p*-value = 0.9559)

8.2.3 *Parametrization of Tree Population*

In this section, we parameterized the effective frontal area density of trees (λ_{f_tree}), as the area-weighted plant area index of a site. λ_{f_tree} depends on the porosity (Leaf Area Index (LAI)), typology (typology ratio (R) between vertical and horizontal area of canopy), and population number (N). LAI is defined as the total single-side leaf area normalized by projected tree canyon area, and is essential for the evaporation, radiation extinction, and water and carbon gas exchange (Breda 2003). We applied the LAI provided by Tan and Sia (2009) to estimate the porosity of plant canopies, in which the indirect measurement by plant canopy analyzer LAI 2000 (Licor, Lincoln, Nebraska, USA) was conducted for the most commonly used landscape trees in tropical area. According to the measured LAI values in the field, the density of tree canopies has been categorized as the dense (Close arrangement and multiple stacking of foliage within the canopy), intermediate (Intermediate between "dense" and "open"), and open canopy (sparse foliage arrangement), as tabulated in Tables 8.2 and 8.3. Since Tan and Sia (2009) conducted indirect measurement, i.e., the radiation measurement, all of canopy elements (e.g., plant stems and leaves) were included in the measurement. Consequently, the measured index represents the entire plant, not only leaves. Since we hypothesized that all of canopy elements contribute on the drag force, this LAI measurement fits the objective of our study.

We defined the effective frontal area density of trees (λ_{f_tree}) as the effective aerodynamic surface area (A_{f_tree}) normalized by the site area (A_{site}), as Eq. 8.7. In this definition, we assumed that leaves would not interfere with each other for airflow, i.e., ideally uniform in behavior (Cionco 1965; Nepf 2012). Due to the flexibility and porosity of plants, trees have a smaller interference area than cubic solid roughness elements, such as buildings (Shao and Yang 2005). To calculate A_{f_tree}, we used the horizontally measured LAI that is an indirect measurement (Tan and Sia 2009). We assumed that the orientation of each leaf was random in the entire tree, i.e., ideally uniform in distribution and geometry (Cionco 1965; Nepf 2012), the horizontally measured index then can be converted to the vertical index. A_{f_tree} can be calculated as Eq. 8.8, where A is the vertical canopy area and can be calculated by Eq. 8.9,

Table 8.2 Values of LAI and canopy area of trees in different categories (Tan and Sia 2009)

Category	Sample of tree	LAI	Canopy area (A_c) according to tree girth (G) (m^2)			
			*G*1	*G*2	*G*3	*G*4
Dense canopy	*Filicium decipiens*	4.0	12	36	80	150
Intermediate canopy	*Tabebuia rosea*	3.0	12	36	80	150
Open canopy	*Peltophorum pterocarpum*	2.0	12	36	80	150

where R is the plant canopy typology ratio between A and A_c. R can be classified into spreading canopy ($R < 1$), spherical canopy ($R \approx 1$), and columnar Canopy ($R > 1$), as shown in Table 8.4. Therefore, based on the definition of LAI, and assuming a random leaf distribution and no interference between leaves, the effective frontal area density of trees $\left(\lambda_{f\_tree}\right)$ can be calculated by Eqs. 8.7–8.9 as (Fig. 8.7):

$$\lambda_{f\_tree} = \frac{nA_{f\_tree}}{A_{site}} \text{ (n: tree population)} \tag{8.7}$$

$$A_{f\_tree} = \text{LAI} \cdot A \tag{8.8}$$

$$A = R \cdot A_c \tag{8.9}$$

At last, substituting Eqs. 8.4 and 8.7 into Eq. 8.2, i.e., parametrizing momentum flux and trees' drag force, we can estimate the averaged wind speed (U_c) in the urban canopy layer that includes both trees and buildings.

8.3 Parametric Study

In this section, we used the new urban canopy model to estimate the averaged wind speed in the urban canopy (U_c) with different urban densities and trees species. Two representative urban areas in Hong Kong with low and high densities were chosen to investigate the effect of urban density and urban trees on U_c. As shown in Fig. 8.8, Sheung Wan was chosen as the high-density urban area $\left(\lambda_{f\,0-15\,m} = 0.45; \lambda_p = 0.45\right)$, and Sha Tin was chosen as the low-density urban area $\left(\lambda_{f\,0-15m} = 0.18; \lambda_p = 0.13\right)$. $\lambda_{f\,0-15m}$ and λ_p were calculated using the methods developed by Ng et al. (2011). It should be noted that due to the skimming flow at urban areas, $\lambda_{f\,0-15m}$ was used in this study to represent $\lambda_{f\_building}$ of buildings (Ng et al. 2011; Yuan et al. 2016). According to the NDVI data in Fig. 8.8, we assumed that there was no tree, but only grass within the test area, and calculated values of

Table 8.3 Examples of tree species with different LAI values (Tan and Sia 2009)

Dense canopy	Intermediate canopy	Open canopy
Filicium decipiens	*Tabebuia rosea*	*Peltophorum pterocarpum*

Table 8.4 Category of trees with different typologies: spreading, square, and columnar canopies

Spreading canopy	Spherical canopy	Columnar canopy
Samanea saman 	*Filicium decipiens* 	*Swietenia macrophylla* 

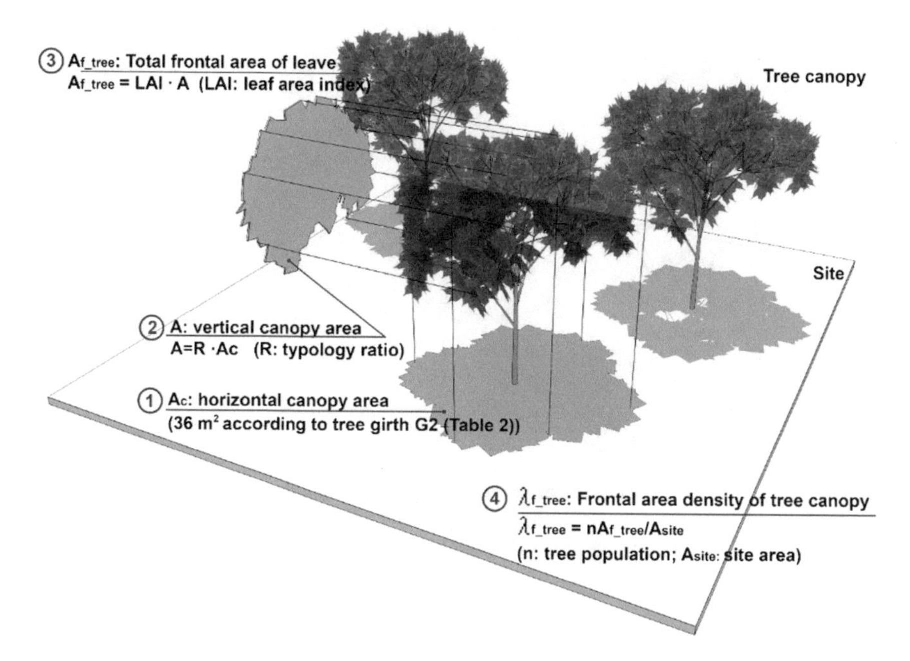

Fig. 8.7 Calculation schematic of effective frontal area density of tree canopy

U_c with different tree species. Two parametric scenarios were modeled with two urban tree parameters, respectively: LAI in Scenario I and R in Scenario II. In both scenarios, the green coverage ratio (λ: the ratio between total horizontal tree canopy area (nA_c) and site area) was increased from 0 to 0.55 and 0.87, which indicated that tree canopies covered all the unbuilt sections, at high- and low-density urban areas, respectively. The parametric modeling results were then cross-compared to evaluate effects of plant canopy density and typologies on U_c.

In this study, we nondimensionalized U_c using U_{ref} as:

$$\mathrm{VR} = \frac{U_c}{U_{ref}}, \tag{8.10}$$

where VR is the velocity ratio, and U_{ref} is the averaged wind velocity at the reference height where there is no building effects on airflow. Based on this definition and Eq. 8.6, VR can be calculated as:

$$\mathrm{VR} = 0.12\left(\frac{2(1-\lambda_p)}{C_{D\_building}\lambda_{f\_building} + C_{D\_tree}\lambda_{f\_tree}}\right)^{0.5}, \tag{8.11}$$

where λ_{f_tree} is the effective frontal area density of tree canopy, and can be calculated using Eqs. 8.7–8.9, as shown in Fig. 8.7. Consequently, VR, as the dependent variable,

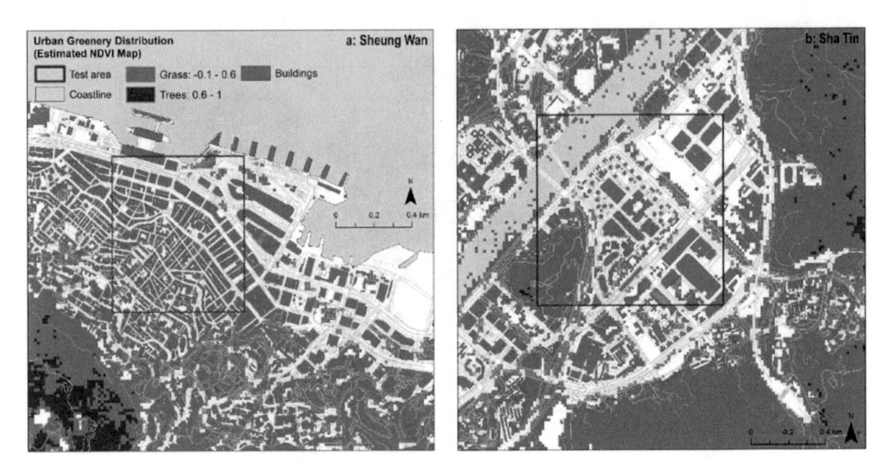

Fig. 8.8 Urban contexts with different densities; **a** Sheung Wan: $\lambda_{f\,0-15m} = 0.45$; $\lambda_p = 0.45$; $\lambda = 0.05$. **b** Sha Tin: $\lambda_{f\,0-15m} = 0.18$; $\lambda_p = 0.13$; $\lambda = 0.25$ (For interpretation of the references to color in the text, the reader is referred to the electronic version of this book)

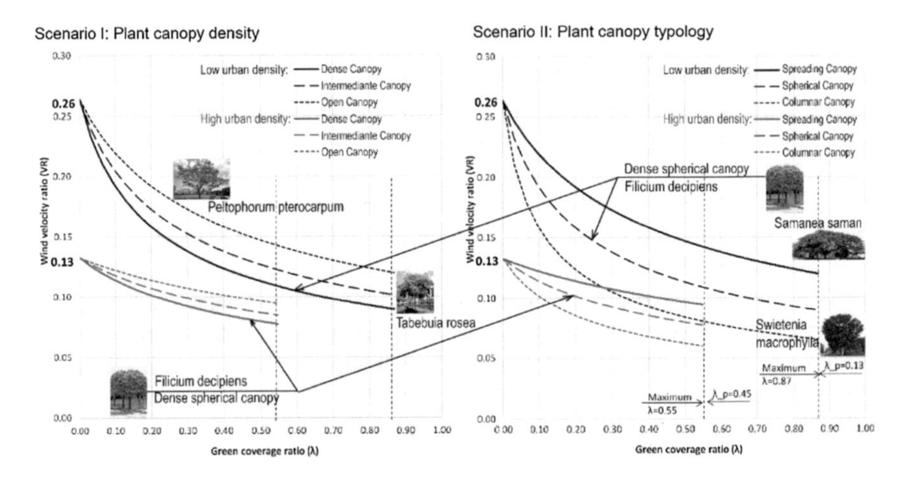

Fig. 8.9 Effects of urban trees on the wind environment at low- and high-density urban areas. Plant canopy densities (Table 8.2) and typologies (Table 8.4), as two parameters, were tested at Scenarios I and II, respectively. Maximum green coverage ratios (λ) are 0.55 and 0.87 in high- and low-density cases, respectively (For interpretation of the references to color in the text, the reader is referred to the electronic version of this book)

can be calculated by green coverage ratio (λ) as the independent variable. The input data (i.e., urban density and urban tree geometry indices) for the parametric study are summarized in Table 8.4. The modeling results are shown in Fig. 8.9.

8.4 Discussion and Implementation

Figure 8.9 summarizes the effects of urban trees on the wind environment at low- and high-density urban areas. Each line type refers to different tree species, and each line color refers to different urban context densities. The figure shows that wind velocity ratio decreases as green coverage ratio increases, i.e., planting more trees at urban areas could slow down airflow at the urban canopy layer, and that the effects of trees on the urban wind environment greatly depend on the density of the urban context, as well as the density and typology of plant canopy.

Specifically, the impact of any particular tree population is different with varying urban densities. As depicted in Fig. 8.9, the VR value decreases more rapidly at the low-density urban areas (black lines) than that at the high-density urban areas (red lines), even though the tree species and number of population are the same. Since the drag forces of both buildings and trees are in the denominator in algorithm, and buildings have much larger drag force coefficient, the impact of trees on the wind speed decreases as the urban area becomes denser. Physically, a high-density urban area induces a weaker wind environment, and that planting more trees does not further decrease the already-slow wind speed.

This observation shows that it is necessary to investigate trees' effects within the urban context, rather than investigating the trees alone without taking into account urban density. For example, if green coverage ratio rises to 40% in low-density urban areas by planting *Filicium decipiens* (Tables 8.3 and 8.4) that has the dense spherical canopy (LAI: 4.0; R: 1.0), the wind velocity ratio could decrease from 0.26 to 0.13 (Fig. 8.9). Such wind environment in the low-density urban area could be similar with the one in the high-density urban areas without trees. Given the wind speed at the reference height is 6.7 m/s, the wind speed could decrease about 0.9 m/s, which can pose significant impact to the outdoor thermal comfort level. In other words, if we plant the same tree population in low-density urban areas, the impact of urban trees is two times larger than the one in high-density urban areas, where wind speed could only decrease by 0.5 m/s. Urban density, represented by $\lambda_{f\,0-15\,m}$, was classified into four categories showing varying effect of trees on urban wind environment (Fig. 8.10). Class "High effect" means that effect of trees is equal or more significant than the black lines in Fig. 8.9; and Class "Low effect" means that effect of trees is equal or lower than the red lines in Fig. 8.9. Furthermore, considering the existing understandings of urban surface roughness and potential air path (Ng et al. 2011; Yuan et al. 2014), it is important to note that only the areas with low-density have been identified as the potential air paths, and thus the impact of urban trees on the performance of these potential air paths is significant.

Moreover, the effect of trees on urban wind environment depends on the tree species, i.e., plant canopy density and typology. As shown in Fig. 8.9 (Scenario I), open canopy trees (short dash lines) have the smallest impact on wind velocity ratio, followed by intermediate canopy (long dash lines). Wind velocity ratio decreases most rapidly with dense canopy (solid lines). Since a denser canopy means larger effective frontal area that causes larger drag force of trees on airflow, this observation

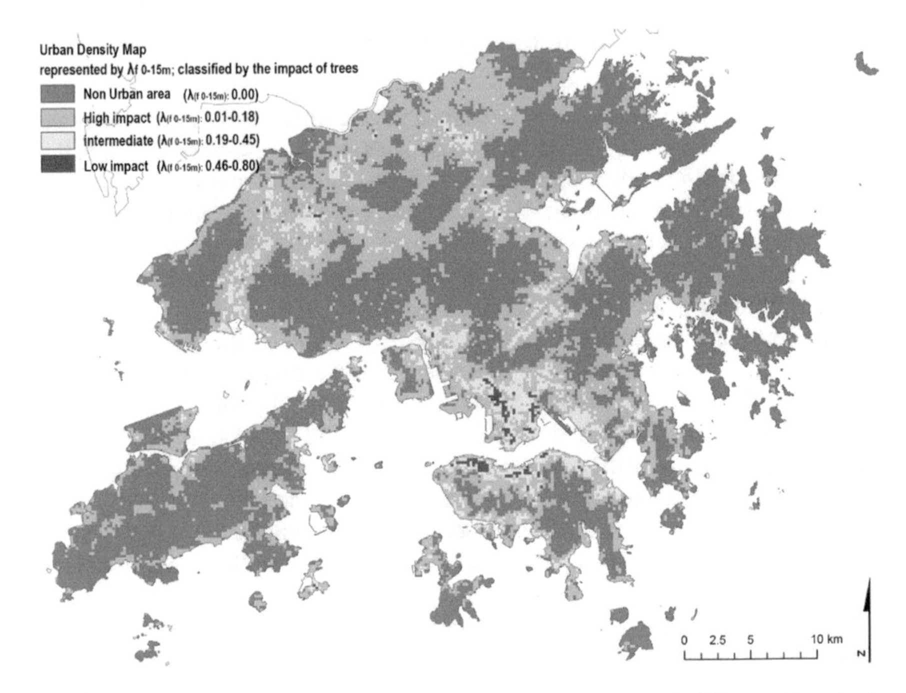

Fig. 8.10 Effect of urban trees on wind environment at urban areas with different urban density (For interpretation of the references to color in the text, the reader is referred to the electronic version of this book)

is considered to be reasonable. In Scenario II, the plant canopy densities are the same, but the typologies are different. The tree species with spreading canopy (solid lines) has the smallest impact on wind velocity ratio. It is because with same green coverage ratio (i.e., same horizontal canopy area), the tree with the spreading canopy has the smallest vertical canopy area, and thus the smallest effective frontal area and drag force.

Such observation indicates that it is still possible to mitigate the negative impact of urban trees by choosing the appropriate tree species, even though the impact of trees on the wind environment at low-density urban areas is significant. For example, we could increase the wind velocity ratio to 0.16 (wind speed: 1.0 m/s) by choosing either *Peltophorum pterocarpum* (open canopy) or *Samanea saman* (rain tree: spreading canopy), rather than planting *F. decipiens* (Dense and spherical canopy) (Fig. 8.9). On the other hand, the wind environment could worsen (short dash line in Scenario II) if we choose an inappropriate tree species, e.g., *Swietenia macrophylla* (dense, columnar canopy). In that case, the wind velocity ratio would be lower than 0.1, and wind speed of <0.6 m/s could lead to significant outdoor thermal discomfort. Suggestions of tree species to address the natural ventilation issues at the urban areas are shown in Fig. 8.11.

Fig. 8.11 Suggestions of tree species to address natural ventilation issues at urban areas

8.5 Conclusion and Future Work

While urban vegetation can mitigate the negative impacts of high-density living (e.g., promote urban biodiversity), it can also slow down airflow or trap air pollutant and anthropogenic heat. In this study, we focused on the negative effect of trees' drag force on the airflow in the urban canopy layer. We developed a new semiempirical urban canopy model that correlates the urban density and tree geometry indices (Table 8.5) with urban wind environment (Fig. 8.8). This semiempirical modeling method was derived based on the balance between momentum flux and drag force of buildings and trees, in which friction velocity and drag force of trees were parametrized. Consistent with our prior research (Ng et al. 2011; Yuan et al. 2014, 2016), the new semiempirical model calculated indices of urban density and tree geometry, without the expensive fluid mechanics calculation. Thus, this new model is considered to be more practical than CFD simulation and wind tunnel experiment, and can provide more direct modeling results to guide the planning and design process.

Moreover, we conducted a parametric study, and demonstrated the effects of urban context density and tree species (i.e., plant canopy density and typologies) on wind environment at high- and low-density urban areas. The new modeling method can provide urban planners and designers both the scientific understandings and practical modeling tool to conduct knowledge-based landscape planning with the goal of introducing more trees into urban areas, and avoiding the negative effects of urban trees on the outdoor wind environment in cities at the same time. Future studies are warranted to integrate the understandings of trees' effect, such as shading and evaporation, on the urban environment. In addition, measurement data from urban- or neighborhood-scale CFD simulation or wind tunnel experiment that incorporates urban trees are needed to further validate and modify our semiempirical model.

Table 8.5 Input data: urban density and urban tree geometry indices

Scenario I	Site	Urban density		Urban tree geometry				
		$\lambda_{(\_f0-15\,m)}$	$\lambda_{\_p}$	A_c	R	LAI (dense canopy)	LAI (intermediate canopy)	LAI (open canopy)
High-density case	Sheung Wan	0.45	0.45	36	1.0	4.0	3.0	2.0
Low-density case	Sha Tin	0.18	0.13	36	1.0	4.0	3.0	2.0
Scenario II	Site	Urban density		Urban tree geometry				
		$\lambda_{(\_f0-15\,m)}$	$\lambda_{\_p}$	A_c	LAI	R (spreading canopy)	R (spherical canopy)	R (columnar canopy)
High-density case	Sheung Wan	0.45	0.45	36	4.0	0.5	1.0	2.0
Low-density case	Sha Tin	0.18	0.13	36	4.0	0.5	1.0	2.0

Two tree geometry parameters in two scenarios are highlighted, i.e., LAI in Scenario I and R in Scenario II

Note

(1) Frontal area density from 0–15 m; $\lambda_{\_p}$ site coverage ratio; A_c plant canopy area (36 m^2) according to tree girth (*G2* in Table 8.2); R plant typology ratio between vertical and horizontal crown projected areas; *LAI* leaf area index

(2) Maximum values of green coverage ratio (λ) in high and low density cases are 0.55 and 0.87, respectively, which means that the sites were totally covered by both buildings and trees

References

Aubrun S, Koppmann R, Leitl B, Mollmann CM, Schaub A (2005) Physical modelling of a complex forest area in a wind tunnel—comparison with field data. Agric For Meteorol 129(3–4):121–135
Bentham T, Britter R (2003) Spatially averaged flow within obstacle arrays. Atmos Environ 37(15):2037–2043
Bot ETG (2014) Turbulence assessment with ground based LiDARs, vol ECN-E—14-043. Energy Research Centre Netherlands
Breda JJN (2003) Ground-based measurements of leaf area index: a review of methods, instruments and current controversies. J Exp Bot 54(392):2403–2417
Britter R, Hanna SR (2003) Flow and dispersion in urban areas. Annu Rev Fluid Mech 35(1):469–496
Bureau of Shanghai World Expo Coordination (2009) Environmental report expo 2010. Shanghai, China
Cheng H, Castro PI (2002) Near wall flow over urban-like roughness. Bound Layer Meteorol 104(2):229–259
Cionco RM (1965) A mathematical model for air flow in a vegetative canopy. J Appl Meteorol Climatol 4(517–522)
Coceal O, Belcher SE (2004) A canopy model of mean winds through urban areas. Q J Roy Meteor Soc 130(599):1349–1372
Cullen S (2005) Trees and wind: a practical consideration of the drag equation velocity exponent for urban tree risk management. J Arboric 31(3):101–113
Dimoudi A, Nikolopoulou M (2003) Vegetation in the urban environment: microclimatic analysis and benefits. Energy Build 35(1):69–76
Gál T, Unger J (2009) Detection of ventilation paths using high-resolution roughness parameter mapping in a large urban area. Build Environ 44(1):198–206
Grimmond CSB, Oke TR (1999) Aerodynamic properties of urban areas derived from analysis of surface form. J. Appl Meteorol 38:1262–1292
Grimmond CSB, King TS, Roth M, Oke TR (1998) Aerodynamic roughness of urban areas derived from wind observations. Bound-Lay Meteorol 89(1):1–24
Gromke C (2011) A vegetation modeling concept for building and environmental aerodynamics wind tunnel tests and its application in pollutant dispersion studies. Environ Pollutant 159:2094–2099
Gromke C, Ruck B (2008) Aerodynamic modelling of trees for small-scale wind tunnel studies. Forestry 81(3):243–248
Gross G (1987) A numerical study of the air flow within and around a single tree. Bound Layer Meteorol 40(311–327)
Hiraoka H (1993) Modelling of turbulent flows within plant/urban canopies. J Wind Eng Ind Aerodyn 46/47:173–182
Hong Kong Civil Engineering and Development Department (2010) Greening master plan. Hong Kong
Hong Kong Planning Department (HKPD) (2005) Feasibility study for establishment of air ventilation assessment system, Final report. The government of the Hong Kong Special Administrative Region
Hong Kong Planning Department (HKPD) (2008) Urban climatic map and standards for wind environment—feasibility study, working paper 2B: wind tunnel benchmarking studies, batch I. The government of the Hong Kong Special Administrative Region
Hong Kong Planning Department (HKPD) (2011) Hong Kong Planning Standards and Guidelines. The government of the Hong Kong Special Administrative Region
Jeanjean APR, Hinchliffe G, Mcmullan WA, Monks PS, Leigh RJ (2015) A CFD study on the effectiveness of trees to disperse road traffic emissions at a city scale. Atmos Environ 120:1–14

Kitagawa K, Iwama S, Fukui S, Sunaoka Y, Yazawa H, Usami A, Naramoto M, Uchida T, Saito S, Mizunaga H (2015) Effects of components of the leaf area distribution on drag relations for Cryptomeria Japonica and *Chamaecyparis obtusa*. Eur J For Res 134:403–414
Kowarik I (2011) Novel urban ecosystems, biodiversity, and conservation. Environ Pollut 159(8–9):1974–1983
Krayenhoff ES, Santiago AJL, Martili A, Christen A, Oke TR (2015) Parametrization of drag and turbulence for urban neighbourhoods with trees. Bound Layer Meteorol 156(2):157–189
Kumer VM, Reuder J, Dorninger M, Zauner R, Grubisic V (2016) Turbulent kinetic energy estimates from profiling wind LiDAR measurements and their potential for wind energy applications. Renew Energy 99:898–910
Lettau H (1969) Note on aerodynamic roughness-parameter estimation on the basis of roughness-element description. J Appl Meteorol 8:828–832
Li XX, Norford LK (2016) Evaluation of cool roof and vegetation in mitigating urban heat island in a tropical city, Singapore. Urban Clim 16:59–74
Loughner CP, Allen DJ, Zhang DL, Pickering KE, Dickerson RR, Landry L (2012) Roles of urban tree canopy and buildings in urban heat island effects: parameterization and preliminary results. J Appl Meteorol Climatol 51:1775–1793
MacDonald RW, Griffiths RF, Hall DJ (1998) An improved method for the estimation of surface roughness of obstacle arrays. Atmos Environ 32(11):1857–1864
Mochida A, Tabata Y, Iwata T, Yoshino H (2008) Examining tree canopy models for CFD prediction of wind environment at pedestrian level. J Wind Eng Ind Aerodyn 96:1667–1677
Nepf HM (2012) Flow and transport in regions with aquatic vegetation. Annu Rev Fluid Mech 44(1):123–142
Ng E, Yuan C, Chen L, Ren C, Fung JCH (2011) Improving the wind environment in high-density cities by understanding urban morphology and surface roughness: a study in Hong Kong. Landscape Urban Plann 101(1):59–74
Ng E, Chen L, Wang Y, Yuan C (2012) A study on the cooling effects of greening in a high-density city: an experience from Hong Kong. Build Environ 47:256–271
Oke TR (1987) Boundary layer climates, 2nd edn. Methuen, Inc., USA
Oke TR (2006) Initial guidance to obtain representative meteorological observations at urban sites. World Meteorological organization
Raupach MR (1992) Drag and drag partition on rough surfaces. Bound-Lay Meteorol 60:375–395
Roth M (2007) Review of urban climate research in (sub)tropical regions. Int J Climatol 27:1859 – 1873
Savard J-PL, Clergeau P, Mennechez G (2000) Biodiversity concepts and urban ecosystems. Landscape Urban Plann 48(3–4):131–142
Schlichting H, Gersten K (2000) Boundary-layer theory, 8th rev. and enl. edn. Springer, Berlin
Shahidan MF, Shariff MKM, Jones P, Salleh E, Abdullah AM (2010) A comparison of *Mesua ferrea* L. and *Hura crepitans* L. for shade creation and radiation modification in improving thermal comfort. Landscape Urban Plann 97(3):168–181
Shao Y, Yang Y (2005) A scheme for drag partition over rough surfaces. Atmos Environ 39 (38):7351–7361
Singapore National Parks Board (2015) Annual Report Singapore Government, Singapore
Tan PY, Sia A (2009) Leaf area index of tropical plants. Centre for Urban Greenery and Ecology; BCA/NPARKS Green Mark for Parks; National Parks, Singapore
Thompson CW, Aspinall P, Roe J (2014) Access to green space in disadvantaged urban communities: evidence of salutogenic effects based on biomarker and self-report measures of wellbeing. Procedia Social Behav Sci 153:10–22
Tokyo Metropolitan Government (2007) Basic policies for the 10-year project for green Tokyo. Tokyo, Japan
Tominaga Y, Mochida A, Yoshie R, Kataoka H, Nozu T, Yoshikawa M, Shirasawa T (2008) AIJ guidelines for practical applications of CFD to pedestrian wind environment around buildings. J Wind Eng Ind Aerodyn 96:1749–1761

Vos P, Maiheu B, Vankerkom J, Janssen S (2013) Improving local air quality in cities: to tree or not to tree? Environ Pollutant 183:113–122

Walker IJ (2005) Physical and logistical considerations of using ultrasonic anemometers in aeolian sediment transport research. Geomorphology 68:57–76

Wong MS, Nichol JE, To PH, Wang J (2010) A simple method for designation of urban ventilation corridors and its application to urban heat island analysis. Build Environ 45(8):1880–1889

Yim SHL, Fung JCH, Lau AKH, Kot SC (2009) Air ventilation impacts of the "wall effect" resulting from the alignment of high-rise buildings. Atmos Environ 43(32):4982–4994

Yuan C, Ren C, Ng E (2014) GIS-based surface roughness evaluation in the urban planning system to improve the wind environment—a study in Wuhan, China. Urban Clim 10:585–593

Yuan C, Norford L, Britter R, Ng E (2016) A modelling-mapping approach for fine-scale assessment of pedestrian-level wind in high-density cities. Build Environ 97:152–165

Index

C. Yuan, *Urban Wind Environment*, SpringerBriefs in Architectural Design and Technology, https://doi.org/10.1007/978-981-10-5451-8

Printed in the United States
By Bookmasters